D0058824

1001 QUESTIONS ANSWERED ABOUT THE SEASHORE

N.J. Berrill and Jacquelyn Berrill

1001 QUESTIONS ANSWERED ABOUT THE SEASHORE

Illustrated with
Photographs and Drawings

Dover Publications, Inc., New York

Copyright © 1957 by N. J. Berrill and Jacquelyn Berrill.
All rights reserved under Pan American and International Copyright Conventions.

Published in Canada by General Publishing Company, Ltd., 30 Lesmill Road, Don Mills, Toronto, Ontario.
Published in the United Kingdom by Contable and Company, Ltd., 10 Orange Street, London WC 2.

This Dover edition, first published in 1976, is an unabridged and unaltered republication of the work originally published by Dodd, Mead & Company, New York, in 1957.

International Standard Book Number: 0-486-23366-9
Library of Congress Catalog Card Number: 76-12889

Manufactured in the United States of America
Dover Publications, Inc.
180 Varick Street
New York, N.Y. 10014

To
Michael, Elsilyn and Peggy
who have shared our
own life on the
seashore

PREFACE

Seashores belong to the sea and almost all life that may be seen on them, apart from birds and humans, comes from the sea. Salt water is tolerated by only a few of the uncountable kinds of insects and not at all by creatures such as frogs and salamanders. For the most part they venture on to the seashore at their peril, in contrast to the creatures and plants of the sea which have colonized the shores in rich profusion. Taken altogether the diversity of marine life along the coasts of the Atlantic and Pacific may well be over- whelming. Only a minority of the animals are obvious and common enough to have been given common names, although those that have not been so blessed are usually just as interesting and merely require a little closer attention or more assiduous search. There is virtually no end to their number and no end to the questions that might be asked about them. Yet somewhere a line has had to be drawn, which is between the reasonably common and the truly rare, and at the spray line above high water and at wading depth below low water. Within these limits we hope to have included most forms of life that are likely to be encountered, and to have answered the questions most likely to arise. At the same time we realize that many aspects of the seashore and its life have certainly been overlooked and that many other questions will naturally arise in the minds of those who use this book. The reader is strongly invited to send to the publishers any questions which may occur to him or to her con- cerning the life of the shore and the shallow seas, so that a subse- quent edition may represent a co-operative venture of those who have discovered the deeper fascination of the sea's margin.

Scientific names have been employed throughout these pages only when necessary for distinguishing one creature from another or for making it possible to refer to the present book from standard identification guides such as Miner's *Field Book of Seashore Life* or Percy Morris' *Field Guide to Shells*. If with the aid of a guide you have found an animal and have only its scientific name, you can refer to the index at the back of this book and so to the questions and answers concerning it. The index, of course, should be employed in the same way when you know the common name, if such there be. To some extent this question book may serve for identifying animals and plants of the seashore, but it is not designed primarily for that purpose. First and foremost it is meant to satisfy your curiosity concerning what you have already found and have a name for, or to tell you where and how to find this or that animal or plant. For quick recognition and naming of the commoner kinds on both the Atlantic and Pacific Coasts, the small colorful guide of Zim and Ingle is strongly recommended and should be used in connection with this book of answers to your questions.

N. J. BERRILL
JACQUELYN BERRILL

CONTENTS

x   **CONTENTS**

PHOTOGRAPHS

(Following page 148)

Plumose anemones and common starfish
Sun star
Sea urchin
Common starfish showing sieve plate and single ray regenerating four
    new rays
Sea urchin shell or test
Brittle or serpent star crawling over colonies of small sea anemones
Rock crab
Green shore crabs mating
Horseshoe crab
Gooseneck barnacle showing stalk, protective shells and fan-like limbs
Male fiddler crab
Limpets in rock crevice
Rock-purple dog whelks feeding on acorn barnacles
Sea snail showing crawling foot, mantle fold around base of shell,
    breathing siphon, tentacles and eye
File shell or swimming clam
Chiton with eight shell plates and marginal fringe
Spiny lobster
Neck of long neck or soft shell clam showing intake and outgoing siphon
    openings
Inner surface of surf clam showing hinge and the marks of the two shell-
    closing muscles and mantle edge
Flatworm
Feather-duster worm showing crown of tentacles protruding from tube

1001 QUESTIONS ANSWERED ABOUT THE SEASHORE

I. THE SEASHORE

1. Why should you be interested in the life of a seashore? In the first place, if you already have this book in your hand you are probably on or near a seashore, or have prospects of going to one, so why waste such an opportunity? If you are on vacation you may even now be a little bit bored with sitting around getting a sun tan and a variety of beach views. In any case the nonhuman life of the shore and the shallow sea beyond it is fantastic, of truly great interest and of almost unbelievable antiquity. It is time you became acquainted, for here, right on the shore at your feet, is as wild a jungle in its own way as any tropical forest and of a kind that may well outlast mankind itself. But the creatures which inhabit it either hide from view or are so still that you accept them as part of the maritime landscape. They are worth getting to know, for their way of life is as successful as yours and has been tested for a much longer time. Every living thing between high tide and low is a colonist who meets his sea of troubles day by day in a quiet but effective manner. Those that you can find will surprise you with both their numbers and their variety. Some of them will be beautiful, some bizarre, but each one fascinating in its own peculiar way.

2. What sort of shore do you have? If you wish to get some feeling of what is going on between you and the sea, look around and first decide what kind of shore lies before you. Is it an exposed rocky shore with pools and ledges all battered by the waves, or a sheltered rocky shore protected from the open sea and on which the tide gently ebbs and flows? Or is it wide stretches of sand with nothing else, or is the sandy beach divided by shallow inlets leading to sheltered mud flats? Are there wharf piles and floats in the cove or harbor, or do you live close to where the mangroves grow? How much does the tide rise and fall, two feet or twenty? So much of what you can expect to find depends on circumstances such as these. For life between the tides must contend with sand or mud or rock, with water or the lack of it, with exposure in various ways to heat and drying winds, and with predators of one sort or another. It is one of the most hazardous regions in the world in which to live, and to understand it all should be both a challenge and a delight.

COLLECTING

3. What are the best ways of finding or collecting marine animals without using a boat? Looking in rock pools or in pools left among sand or mud flats. Lifting the curtain of seaweed draped over rocks, especially near the level of low tide. Turning over boulders along the shore—always returning them to their original position—the most promising being those at or just below the low-tide level that have a little space between the underside and the ground. And, not least, by digging with a spade or fork in sand or sandy mud at the edge of the water, particularly at low tide. Possibly the most rewarding and most ridiculous-looking procedure is to lie down on the wet, weed-covered rocks at low tide with your feet higher than your head and your face in a crevice or beneath an overhanging ledge, the closer to the water the better. A wave may come up and douse you or you may even slip in, but the experience is usually worthwhile and the feel of seaweed draped over your neck with nothing of the human world in sight is a memorable one. (See question 15.)

4. How can you find animals that live on a sandy shore? Most creatures on a sandy shore lie buried beneath the surface, at least when the tide is out. Some lie just beneath the surface, if that, while others burrow deep. You need a keen pair of eyes and a spade in order to find them.

If the sand is barely exposed and almost flush with water, as it is near the low-tide level along the great beaches around Cape Cod and Cape Hatteras, a number of animals can be trailed. A short track that ends in a small mound probably indicates a moon snail, which you can lift out by digging your finger in the sand at the edge of the mound. A broad, short trail that has no mound at the end is likely to yield a keyhole sand dollar, while long meandering trails most often end with a small horseshoe crab. (See questions 224, 592, 666.)

Apart from trails some of the burrowing animals also leave indications of their presence at the surface. The lugworms and acorn worms leave casts of mud or sand at the surface at one end of their burrows, which are certain giveaways, while clams of various kinds often leave small distinctive holes marking the positions of their siphons, particularly where the sand or mud is both firm and wet. Also when a clam suddenly withdraws into its burrow it usually sends a spit of water up into the air. But if you wish to find any such creatures as these you mark the spot if possible and just dig with a spade for as long as your strength or interest holds out. (See questions 783, 903, 391.)

5. How can you best find animals on a rocky shore? By examining the rock pools from the high-tide level down to the lowest possible. At the lower level the pools which are deepest and have the most seaweed are likely to contain the greatest variety of animals. You may need to wade into them, in spite of the great unknown, or to lie down at the edge and explore blindly along the bottom and sides with your hand. The bottom may yield sea urchins, starfish and the occasional startled and startling crab. The sides may be draped with all kinds of encrusting or otherwise attached animal growths, while kelp pulled up from the bottom will usually have many creatures entangled in its holdfasts. Others again crawl on or live attached to the blades of the weed.

Apart from the pools you can find a number of animals by lifting up the curtains of seaweed covering the lower rocks, while the undersides of overhanging ledges very close to the level of low tide are generally most rewarding, for here you will find colorful anemones and carpets of green and red sponges hanging down toward the water. Turning over boulders is a pastime in itself that yields many surprises. (See questions 13, 14, 76, 505, 801, 809, 893, 895, 911.)

6. What do you need for observing the marine life below the level of low tide? If you wish to see the marine life in shallow water rather than collect it, several methods are possible. The older way is to go out in a skiff, preferably with someone who will row you slowly around, and look at the sea floor over the side of the boat with a water glass, that is, a wooden box with a glass bottom. The glass gives you a window to see through and the sides of the box cut out the reflections from the sky.

Alternatively you can wear goggles and mask and either go wading up to your chest or better still swim or float face down at the surface, using a snorkel for breathing and flippers to aid your progress. Even in very shallow water, just deep enough to float in, lying face down with mask and snorkel can be an education in itself if the marine life is at all abundant. You get the feeling of having entered a different world.

7. How can you prevent the inside surface of goggles or mask from clouding? Spit in it and rub the surface before wetting it with seawater.

8. What do you need to study the marine life of a rocky shore? A pocket or hand lens of some kind, since so much of the marine life is either small or has fine detail that can be appreciated only with some magnification. A glass jar is also desirable, into which you can put the smaller specimens together with some water in order to examine them better. Since clambering over slippery rocks with breakable glass in your hand is hazardous, the glass is better carried in a wooden or tin container. If you are interested in some of the finer things of life, a pair of tweezers will be useful for detaching them from rocks and weeds. Sneakers are desirable since the barnacles on a rocky shore can too readily cut your feet.

9. What are the dangers of exploring rocky shores? Apart from the possible approach of unusually high waves on the more northern Pacific Coast (see question 15) there is little or no danger if you have a companion. Otherwise there is always the possibility that you may turn an ankle or break a leg or get your foot jammed in a crevice, any one of which may cost you your life if no one is around or if your voice cannot be heard above the sound of the surf or the crying of seagulls.

10. What can you expect to find in a rock pool near the high-tide level? Mainly those animals that cannot live long out of water but are at the same time capable of tolerating the increase in temperature which usually occurs in high-level pools when the tide recedes. The green seaweed or "sea hair" (*Enteromorpha*) generally lines such pools, particularly where there is some seepage of fresh water. The commoner animals are small snails, limpets and side-swimming crustaceans.

11. What animals are likely to be found on exposed rocky shores above the mid-tide level? This is the region for specialists designed to withstand exposure to surf and sun. The predominant ones are certain periwinkles, limpets and, above all, the acorn barnacles. Some mussels are also often abundant where there is at least some semblance of a rock pool. (See questions 5, 18, 13.)

12. What animals live on the exposed rocks below the mid-tide level? The acorn-barnacle zone extends well down into this region, but where the surf is too strong for seaweeds to settle, the mussels for the most part take over and form blue-black carpets of living bivalves. On the Pacific Coast, but not on the Atlantic, a similar carpeting of the green sea anemones is also commonplace. Where the rocks are draped with weeds, however, all kinds of animals find shelter from sun and wind and their names are legion.

13. What can you expect to find on the underside of rock boulders? If the boulder is the right sort with regard to its position on the shore (see question 14) its undersurface is likely to be encrusted with a variety of animal life and plant growths, the plants being mainly of the encrusting coralline kind (see question 113). The encrusting animal growth will be principally sea mats, sponges and colonial sea squirts (see questions 894, 875, 909), while attached but pendant forms may be either sea mats of another kind or hydroids (question 804) or sponges again, or all three. Solitary sea squirts and sea anemones also may be attached to the undersurface if there is space enough, together with tube worms, chitons and limpets, not to mention various flatworms, sea slugs and brittle stars that may be browsing on the helpless animal growths. (See questions 742, 447, 685.)

The shallow pool below the rock is also a haven in which crabs and shore fish may shelter until the tide returns, while the ground beneath is a good place to dig for burrowing shrimps and worms.

14. What rocks or boulders are the most worthwhile turning over, and why should a disturbed boulder be replaced? The larger the

boulder the more likely you are to find the territory beneath well occupied, since stability of the rock is important to the under-rock community. Consequently a crowbar and a strong back are valuable
assets. If a boulder is well bedded in sand or mud, however, you may
find little for your efforts. The most profitable are those near the level
of low tide which have enough space between the undersurface and
the ground for water to circulate between and for plant and animal
growths to suspend themselves from the rocky roof.

All disturbed boulders should be replaced because everything that
grows attached to their undersides or lives beneath them will die if
left exposed.

**15. What should you do if you are low down on a rocky shore and
see a large wave coming?** On the Atlantic Coast occasional waves
of unusual height are very rare except after a storm when heavy surf
is usually breaking over the rocks. Under such circumstances you
should not be within reach of the waves at all and the scene should
merely be viewed from a respectful distance. In any case do not under-
estimate the force of even small waves surging over rocks. On the
Pacific Coast the situation is more subtly dangerous, for ground swells
sweep in from the ocean, caused by some distant storm, when no immediate reason for alarm is evident. Exposed rocky points from central California northward are very dangerous places. According to
E. K. Ricketts and Jack Calvin, "any person within twenty feet of the
water, vertically, is in constant danger of losing his life, and every year
the newspapers report, with monotonous regularity, the death of people who have been swept from the rocks by unexpected waves of
great size. Such loss of life is usually unnecessary as it is regrettable.
To lie down and cling to the rocks like a starfish and let a great wave
pour over one takes nerve and a cool head, but more often than not
it is the only course of action. To run, unless the distance is very short,
is likely to be fatal."

**16. What animals are most likely to be found on a sandy or muddy
shore?** Animals of two or three kinds, so far as their way of life is
concerned. One group lives in the sand or mud, simply using it as an
anchorage in place of stones and rock. Certain sponges, sea pens and
tube-building worms are in this category, all equipped to hold their
place and yet raise themselves above the surface. Other forms crawl

on the sandy or muddy surface, or no more than an inch or so below; for example, horseshoe crabs, green crabs, fiddler crabs and others, certain snails, sand dollars and even a few fish. The third group is the largest and consists of the burrowers, which include a host of clams, worms and burrowing shrimps, etc.

17. What are the dangers of collecting on sandy or muddy shores? If the rise and fall of the tide is not very great, the danger is slight, but wherever the tide range is considerable there is often danger that you may wander out with the edge of the receding tide, wading through intervening channels where the water remains, and fail to notice in time that the tide has long since turned and left you marooned on a sand bar soon to be submerged, with too long a distance and here and there too deep water for a safe return to dry land.

18. How do you find animals that live on mud flats? Find telltale marks on the surface if possible and in any case you have to dig, dig, dig, either with a fork or a clam rake. Rubber boots may keep your legs from getting too muddy, so long as you do not get in too deep, though many a boot has been left behind in the mud because the owner could not withdraw his foot with the boot still on. Bare feet are better and you will soon get a liking for the sticky squashiness of the muck as your feet and legs sink into it. Any small child would revel in it and this is probably your last excuse to behave as such. (See questions 391, 399, 402, 404.)

19. What animals are likely to be found in pools among mud and sand flats at low tide? Various snails, hermit crabs, shrimps, small flounders and gobies, and an occasional rarity of almost any kind. (See questions 259, 286, 617, 592, 964, 999.)

20. What plants are found along the beaches and sand dunes that help keep the sand in place? Where the blown sand slows up sufficiently for plants to take root, several kinds soon get started and once established take hold of the sand and keep it in place. Above all others, outside of the tropics, a beach grass known as marron grass or shore rush performs this function. Like all rushes it spreads by means of underground runners, sprouting through the sand at more or less regular intervals, so that before long what was a shifting dune of

Beech oats (2–3′)

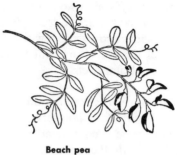

Beach pea
(flower spike 1–2″)

Dune lupin (1–2′)

Railroad vine (4–6″)

Ice plant (4–5″)

Bearberry (4–5″)

Wormwood (12″)

Hudsonia (3″)

Fig marigold (2")

Bayberry (3–4')

Salicornia (8")

Echeveria of the
West Coast dunes

Marram Grass (12–15")

blowing sand looks like a sparsely sown field of young wheat about a foot high. Gradually the spaces between fill up until the roots and runners link together beneath the surface in a tough, binding network. Once this has happened, other plants have a chance to grow—other grasses such as sea oats, sand burs and shorter turf-making kinds. Then the bushier plants find their place, especially where there is gravel or shingle as well as sand, so that the beach margin and the rear of sand dunes harbor beach plum, beach pea, wild rose, cat briar, bayberry and, usually farther in from the beach, straggling groves of scrub pine.

Along more southerly beaches the process of the sand mastery by the land plants is aided by the long tough large-leaved vine known as the sea grapevine or railroad vine, which sends its runners down the seaward slopes to the edge of the sea itself. While in the semitropics and the full tropics the green, fleshy, juicy-leaved, sesuvium carpets the sand a little above the high-tide level, and coconuts and mangrove seedlings, tossed by wind and wave above the line, take deep root and establish a barricade and a green cover that holds the sea at bay and gives protection to a host of other plants and animals.

Beach plants have a lot in common with desert plants for both have to contend with dry sand and full exposure to sun and wind, although beach plants have salt spray to withstand as well. Mostly they tend to be unusually woody or exceptionally juicy, either to cut down their need to hold water or to hold on to water with remarkable tenacity. The sesuvium of the Keys is cactus-like in its fleshy juiciness, and so are the leaves and stems of the sea figs (fig marigolds), the ice plant and the lovely Echeveria of the West Coast. The latter are readily recognized. The Echeveria has long, thick pointed leaves growing as a rosette six or eight inches high, with two-foot spikes of dark pink buds which open to show starry yellow interiors. The ice plant has beautiful, big, wavy dark green leaves, tinged with red and studded with crystal dots that look like ice. While the sea figs have narrow but thick fleshy leaves and lavender colored marigold-like flowers. The sand verbenas and seaside daisies of the West Coast are also colorful, the former with leaves that are thick, flat and sticky, lying flat on the ground, with yellow, red or white flower heads standing a few inches high. Possibly the loveliest plant of the western dunes is the dune lupine, a small kind with dainty gray leaves and lavender flowers, but with ten-foot roots wandering over a large surface in their search for water. Beach shrubs are for the most part tough and woody, with hard-surfaced glossy or waxed leaves, as in the case of the bayberry and beach plum and the yaupon, which is a red-berried dune holly of the Atlantic and Gulf Coasts.

21. What is sand? Most of the sand on the seashore comes from the weathering of rocks along the coastal regions, either directly through the action of wind and wave or transported by rivers and rain from more inland regions. With every storm and every tide the material is sifted and sorted and resifted so that the beaches everywhere are continually being remade. As the grinding process goes on the particles become smaller and separate into pieces of various size, each size drifting with wind and water to its distinctive place. Most beach sand consists mainly of crushed quartz, the commonest substance in rocks of almost every kind.

22. Why is some sand coarse and some fine? The coarseness and fineness of a sandy beach depends on the nature of the sea currents and waves which sweep across it. Small, light particles are carried far; heavier, larger particles sink more readily, so that a steady sorting is forever going on, leaving some parts of a beach with coarse sand only and others, particularly higher up the beach, covered deep with sand that is much finer.

23. What gives sand its colors? The rock material from which most sands are made consists of quartz with many other minerals. Some are heavier than others and of different colors, and the sorting action of waves and currents distributes them in various ways, so that some sands are yellow, some greenish, some almost a pale purple, some red, and so on. In places the sand may be virtually black where volcanic rock has been pulverized. In others the contribution from fragmented shells and coral may be large, and the sand is soft, almost chalky, as among the Florida Keys and elsewhere. Every beach and every kind of sand is a sort of ground-up history of part of the earth's crust going back through time immemorial.

24. What makes sand dunes? High tides toss sand to their highest level. On shore, winds carry the higher sand even higher, piling layer upon layer. In certain regions where the sand is inexhaustible, the winds strong and frequent, and no coastal cliff to block progress, the sand is blown and rolled inland year after year to pile up in billowing sand hills fifty feet high. A slow gentle slope usually rises from the beach until you reach a sharp escarpment where the freshly blown sand is forever falling over. In a live sand dune of this sort the hill of sand slowly advances inland like a glacier, engulfing all vegetation and even trees that stand in the way.

25. How are mud banks and salt marshes made? Mud, along inshore coastal regions at least, is mainly the topsoil of the land brought down by the rivers. As soon as the rivers widen into estuaries the flow of water slackens and the fine particles of soil settle slowly on the bottom, chiefly at the sides where the flow is less and particularly in inlets and quiet bays near the mouths of rivers. As more and more is brought down, the banks along a salt creek grow and cause the channel to narrow, until at last sufficient mud accumulates to form a soggy salt marsh throughout the upper part of the creek or bay, through which the fresh-water channel now meanders as it cuts its way to the sea. The salt marsh formed in this manner has its own special forms of life. The mud is a rich soil, but it has sea salt mixed with it from the time it first settled down, while salt is replenished with every high spring tide. Few plants can tolerate so much salt, but one in particular, marsh grass, grows abundantly in the salt mud, converting the flats into salty meadows. Where the salt is not too much, tall marsh reeds, spreading by fast-growing rootstocks, may form dense stands.

Both land and sea creatures find a place in this salty margin combining the two elements. Snails are abundant (see question 286). The commonest is the salt-marsh snail, which ranges from Nova Scotia to Texas. It lives in marshes that are occasionally overflowed by the tide and is never very far from the high tide mark. When the sea flows over the marsh the snails climb to the top of the marsh grass to avoid submersion as long as possible. In southern Florida the coffee-bean shell, which lives on the mud flats, also climbs high on the grasses and shrubs when the tide is full. Fiddler crabs are also common, one species being abundant in the mud and sand of salt marshes close to the sea and another living far up in the marshes, into water that is only slightly brackish. Salt marshes accordingly are happy hunting grounds for ducks in migration and other birds, including wild geese in certain regions.

26. What grows on wharf pilings? A wharf pile is somewhat like a vertical section of a rocky shore, from high-tide level to below low tide. Barnacles and mussels are typical of the upper part exposed for the longest time out of water. Tube worms, tube mollusks, and small anemones may be found lower down, together with encrusting and branching sea mats. Where mussels are common, starfish are most likely to be present as well. Wharf piles standing in clean, quiet water where even at low tide several feet of piling remains below the surface as a rule, are virtual gold mines to the marine naturalist if he can get down that far. Here above all you will find anemones, hydroids, sponges, sea squirts and many other forms growing in wild confusion, a rich community in itself and surpassed only by rock pools close to the low tide level. (See questions 657, 428, 755, 292, 799, 804, 874, 909.)

27. How quickly can a ship bottom or float become covered with marine animals and plants? In moderately warm waters only a few weeks are necessary if the surfaces have not been treated with anti-fouling paint or other protection. In colder waters two or three months may be required before the growth is significant. (See questions 29, 30.)

28. How can you best grow an interesting collection of attached animals and plants? If early in the season, as early as possible during spring or summer, you fasten flat pieces of wood so that they always have the same side downward in the sea and cannot float away, by the end of two months or so you will have a rich growth of weed along the edges and a variety of sea mats, hydroids and encrusting sea squirts on the shaded side. Brought inshore and examined in water with a hand lens, the collection should prove fascinating. (See questions 54–57.)

29. What are the fouling organisms on the undersides of ships and floats? Principally they are the same organisms that commonly encrust the undersides of rocks, namely sea mats, hydroids, sea squirts and sponges. Mussels and green seaweed grow along the sides just below the water line where the light is stronger. (See questions 894, 804, 909, 874.)

30. How can you protect submerged wood from fouling organisms? Copper-containing paint serves to protect submerged wood such as boat bottoms from season to season. Copper sheathing is more durable but far more expensive. Anti-fouling paints of various kinds however are available, while research goes on continually to improve them. The boring gribble and the shipworm, which actually eat into the wood, present much graver problems. (See questions 639, 432.)

31. What grows attached to mangrove roots? Coon oysters, periwinkles, sponges, both solitary and compound sea squirts of many kinds, feather-duster worms, and many smaller forms of life.

32. What lives among the roots of the mangrove? Fiddler crabs by the thousand. Crown conchs that prey upon the coon oysters attached to the mangrove roots. The coffee-bean snail which hesitates between land and sea, and purple-clawed hermit crabs. (See questions 547, 275, 286, 617.)

33. What are the prop-like roots of the mangrove? While serving as roots to some extent their chief purpose seems to be to prop the main stem or trunk against the uprooting pressure of wind and wave, so that a mangrove forest becomes an impenetrable jungle as high as a man, no matter how tall and distinct the mangrove tree trunks stand.

34. Where do mangroves grow? Along the more sheltered shores of tropical and semitropical seas.

35. How do mangroves start to grow on the shore? The mangrove drops its heavy six-inch-long seedlings into the tide to drift away in the sea, to lodge on the shore many miles away or perhaps only a few

Shoot of Mangrove, with seedling dart
(6–8″) about to drop

feet. Wherever they lodge they tend to become wedged in the mud or coral gravel in an upright position, when they send roots down and leaf-shoots up. Buttress roots spring out and down from the sides and soon a well-anchored mangrove has become established.

36. How do mangroves help to make islands? As soon as a group of young mangroves have anchored themselves on an offshore sand bar and sent out their arching circles of supporting props, all manner of things come to rest among the roots, such as coral fragments, decayed vegetation, and many others. Before you know it a small island has been born, on the miniature shores of which other mangrove seedlings lodge and among which more debris accumulates—and so the island grows.

37. What can you see living among turtle grass? One of the most spectacular marine communities in existence. The great conchs such as the queen conch and the fighting conch are here found alive, together with tulip band shells, cask shells and helmet shells in their inhabited state. Giant starfish, sea-biscuits, seahorses and pipefish, small cowfish, green spider crabs, sea hares and young octopi all wander among the turtle-grass forest. Even the sea turtles visit the beds at times in their search for sea biscuits and conchs. (See questions 276, 277, 283, 192, 220, 929, 937, 356, 45–48.)

38. What animals live among or attached to eelgrass? A somewhat similar community to that found among turtle grass (see question 37) although not so spectacular. Scallops rest on the eelgrass floor, ghost shrimps burrow among the roots, hermit crabs and whelks crawl at the bottom, broken-back shrimps and pipefish live among the ribbonlike blades, snails and limpets crawl on them, and stalked jellyfish live attached to them, all well fitted to their particular way of life. (See questions 337, 610, 617, 270, 599, 937, 311, 831.)

39. What animals and plants build reefs? In warm shallow seas the principal reef builders are the stony, hard corals which can construct massive reefs without aid from other forms. In many places, however, certain tube worms add much support to the reef walls with their own layers of limy tubes. Some tube worms form reefs of a sort even by themselves. The tube snail also adds its quota here and there. All of the above are animal. The great bulk of the coral reef, however, consists not only of true coral but of nullipores as well, nullipores being massive stony plants related to the seaweeds. (See questions 44, 840, 761, 292.)

40. What animals do you see crawling on or among coral reefs? There are many kinds. Sea hares, conchs, flamingo-tongue snails, sea urchins, starfish, basket stars, brittle stars, bristle worms, mantis shrimps, spiny lobsters and a host of lesser forms. (See questions 356, 274, 202, 156, 193, 714, 614, 585.)

41. What fish do you commonly see among coral reefs? Many kinds of fish live among or around coral reefs, taking advantage of the shelter the coral crevices and curtains afford, and feeding on the variety of plants and animals the corals support. To describe them all would take us too far offshore and only a brief listing is possible here.

The black angels rank high, black with some yellow mottling when fully grown, black with broad yellow bands when young. Blue, green, yellow or red parrot fish, with overly large scales, whose teeth leave grazing marks on the rocks where they have been scraping off the film of vegetation, compete with the angels for your attention. Smaller but more numerous are the blueheads, red squirrels, serjeant majors, yellow grunts, slippery dicks, and the little butterfly fish with its beauty spot at the base of its tail. Seahorses and the curious box-like cowfish belong here too, and so do the vicious moray eels that remain much of their time with only a wicked-looked head projecting from a coral hole. Gars jump at the surface of the water like flying silvery arrows, over the coral itself and in the lagoons behind, while gray snappers hunt along the edge of the reef in schools, waiting to take their smaller, more colorful brethren by surprise or content to snap up careless crabs or shrimps. Now and then, prowling and alone and moving along with imperceptible disturbance, barracuda, small or large, watch for a meal and should be watched in turn. (See questions 929, 988, 991.)

42. What are the colorful growths you see on coral reefs? Most of them are animal by nature. The stony corals that form the reef supply anchorage for a host of other kinds. The soft corals such as the sea fans, sea rods and sea whips add color and pattern to the rest, and on these in turn you may see enveloping orange or purple colonies of sea squirts. Sponges of various colors—green, red, yellow, etc.— decorate the undersides of coral boulders, while vivid green fleshy seaweeds form a herbaceous edge to the garden. And here and there protruding from small holes and crevices are the bright orange, red and purple heads of the flowering or feather-duster worms. (See questions 841, 853, 856, 921, 881, 754.)

43. Is it safe to touch corals? Many corals can sting like jellyfish, at least intensely enough to be best not handled unnecessarily. (See questions 809, 828.)

44. Why are coral reefs found only in warm seas? Corals as such can grow in relatively cold and deep northern waters, but the particular kinds which contribute to the formation of massive coral reefs are typical of semitropical species well adapted to living close to the maximum temperatures of the sea, where light is usually intense and growth rates can be high. From the nature of things they cannot at the same time survive under very different conditions. The corals that do grow elsewhere are successful as corals but not as reef-builders. (See questions 845, 852.)

45. Why is there usually a shallow lagoon on the land side of a coral reef? A coral reef grows upward to the level of low tide, after which the conditions for continuous active growth are much better on the exposed seaward side than on the other. Surf and storm breaking on the outward side however carry coral rocks and sand over the reef and deposit them on the sheltered side, so that the region between the reef and the shore slowly fills up with trapped material to form a progressively shallow lagoon sheltered from the open sea by the natural coral breakwater.

46. Can you observe marine life at night? Yes. With the aid of a good light—flashlight or other kind—you may see a surprising amount of activity both on the beaches and in the water beside a wharf. Crabs of many kinds roam the beaches at night, while a light held for a time over the edge of a wharf, particularly shortly after nightfall, may attract many kinds of swimming creatures, especially worms in a nuptial mood. Turn off the light, however, and splash the water and you may see the star phosphorescence and possibly other kinds. (See questions 46, 59, 722, 731.)

47. How can marine life best be seen while swimming? With the aid of goggles or mask, and a snorkel. (See questions 50, 1106.)

48. How can you collect marine animals on the sea floor below wading depth? By several means. The most precise and selective method is to swim down equipped with an aqualung (see question 51). An alternative when the water is from six to eight feet deep is to employ a long-handled net from a skiff, preferably a net with a heavy, straight-edged frame. Professional biologists often use some form of small dredge consisting of a heavy rectangular metal frame to

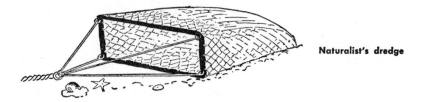

Naturalist's dredge

which is attached a sack or rope bag, the dredge usually being pulled along the sea floor with a long rope operated from a slow-moving power boat. If you enjoy exercise you can work the dredge from a rowboat or skiff. In any event it is a fascinating procedure, like dipping in a grab bag, for you never know what may come up until you see the contents.

49. How can marine animals be seen when using a skiff? With the aid of a water glass, which is a deep wooden box or nail keg with a glass bottom (see question 6) held over the side of the boat. It cuts out the ripples and reflections at the surface and is like looking through a porthole.

50. How can a skin diver best study marine life? Wherever the water is moderately warm and fairly clear, but especially in the south where marine animals are in general more colorful and spectacular, all you need to do is to paddle along with flippers, or merely float at the surface, looking down at the bottom through your mask and breathing quietly through your snorkel. Every so often something entrancing will swim or crawl within view and you feel you share the same watery world with it. If you put your mind to it you can easily get the feeling of being a part of the marine life yourself, and it is an unforgettable experience.

51. Is it necessary to use an aqualung? If you wish to collect sea creatures ten feet or more below the surface or really get down to study them, an aqualung is necessary, for you cannot hold your breath long enough for this purpose. But you must be thoroughly familiar with the equipment and its use, for there have already been far too many fatal mishaps resulting from mechanical failure or improper usage. (See question 1106.)

52. Can skin divers affect marine life? Where skin diving, with or without the use of an aqualung, is popular and is conducted mainly in connection with spear-fishing, the marine life suffers. The fish in particular more or less disappear, partly because of the slaughter, partly because they are intelligent enough to learn from experience and to withdraw to deeper water. The fringe of the sea, with all its delicate living beauty, is more tenuous and fragile than is commonly believed and is decidedly vulnerable to human depredation. The marine life of the Mediterranean coast of France, for instance, has already been decimated and wardens have been on duty for several years to control the enthusiasm of spear-fishers. In any case you can get far more pleasure from following an angel fish and observing its satisfactions, alarms and general behavior, than from any attempt to stick a spear through its body. If you have an eye for beauty, give both yourself and your victim a chance to enjoy life.

53. How can you best examine animals you have collected without taking them away from the shore? Find a small, shallow pool among the rocks or sand and use it as a temporary aquarium. The water in the pool should not be noticeably warmer than the place where you found the animals. Then lie down and watch the animals more or less face to face until you become fully acquainted. A large reading glass will add greatly to your enjoyment and education since the more interesting detail is usually somewhat small.

54. How long will sea creatures live if you take them home in a bucket of water? If all you have is a bucket of seawater, how long the animals live depends on how many there are and how large they are and how cold the water is. If the water can be kept cool, as cool as the place where the creatures came from, and if the animals are few and small, they may live for days in a bucket of water. But if you have not restrained your collector's appetite and have a number of fairly large creatures in your bucket, and if, as is generally the case, you are carrying the bucket of water from the cool sea to a warm location on land, the chances are that all will be dead by the time you arrive. If you have a salt-water aquarium waiting for your animals it is better to transport them in a bucket which has a little water and a lot of sea-weed, which takes care of the oxygen supply better. (See question 855.)

55. What animals can best be kept in a marine aquarium? The first rule is not to have too many or too large, for the greater the amount of water relative to the total quantity of life, the healthier the animals will be. As long as all are healthy, most kinds can be kept together, although starfish are predisposed to attack and devour clams and mussels, while crabs, particularly spider crabs, are adept at cutting off the upper part of a sea urchin shell in order to get at the flesh inside. Otherwise you can keep sea anemones, crabs, hermit crabs, small starfish, scallops, sea snails and small fish more or less happily together in the same tank.

56. Should you keep seaweed in a marine aquarium? Only for the purpose of providing shelter or shade for the animals. As a rule sea-weed soon begins to decay when kept in aquaria and may foul the water. If the aquarium is maintained with a running supply of sea-water and the temperature is not too high, weeds already attached to rocks will survive well enough and give a more natural appearance to the whole.

57. What kind of aquarium can you set up that will take care of marine animals? It depends a great deal on where you are located. If you are fortunate enough, in this particular connection only, to live close to the sea in a region where the air temperature differs little from that of the sea, the greatest problem does not exist. You need only to have a glass aquarium and an efficient aerator such as is commonly used for freshwater aquaria. The one point to remember is that seawater is much more corrosive than fresh water and no metal must be exposed on the inside of the aquarium, for a trace of it dissolved in the water will be highly poisonous to most kinds of marine animals. If you do not have a regular aquarium with an aerator but have large glass vessels such as goldfish bowls, there are still several possibilities. If the temperature can be kept cool you can partly fill the bowls with seawater to a depth of about three inches and keep a variety of small creatures in it for some time. The water should be changed once or twice a day. If the temperature is relatively warm, you can keep only those creatures that are used to such temperatures, which are mainly the animals found along the higher levels of the shore, such as fiddler crabs, rock snails, etc. For these there should be some water but also some sand and rock which stands higher than the water so that the animals can get out when they need to.

The best arrangement, for those that can manage it, is an aquarium set up close to salt water, with an electrically run water pump maintaining a continual flow of seawater through pipes into the aquarium. Under these circumstances almost any kind of marine life can be kept alive for an indefinite period. Aquaria kept outside should be shaded on top.

58. How should animals in a marine aquarium be fed? Crabs, carnivorous snails, sea anemones, etc. relish the flesh of mussels and most other creatures. Crabs and their relatives do not as a rule recognize the presence of meat unless it is placed under the body just behind the claws. In the case of anemones it must be placed or dropped into the middle of the crown of tentacles where the mouth lies. Most marine creatures can do without food for considerable periods, and feeding them in a temporary aquarium serves more to educate the onlooker than to bring health and happiness to the inhabitants. As a matter of fact, soft-bodied animals such as sea anemones, sea cucumbers or flatworms, which have no hard skeleton or shell, can starve without harm, merely becoming smaller and in a real sense recovering their youth.

59. When are you most likely to see action in a marine aquarium? During the night. The light in an aquarium during the day is much brighter than that to which the animals are normally accustomed on the sea floor. Sea anemones contract in a strong light and are fully expanded only during darkness, and the same is true of corals. If you really wish to see how small marine creatures conduct themselves, go look at your aquarium at night with a flashlight. (See questions 46, 843.)

60. Can you photograph marine animals? Photographing marine animals or plants is essentially like close-up portrait photography of faces or flowers. If you have the equipment and experience to do the latter you will have little trouble with the former. In any case, a single-lens reflex camera with either accessory close-up lenses or extension tubes is the most useful and the simplest to use. If the animal can be photographed on rocks, weed or sand out of water without detriment to its lively appearance, sharper pictures can be obtained that way than through water. Otherwise, satisfactory marine-life photography depends on setting up small glass aquaria with suitable background, the important points being to avoid reflections from the glass surface and to be able to focus directly on the animal through the water, for the refractive index of water is such that the focal distance from lens to object cannot be determined by measurement; hence the need of the single-lens reflex type of camera.

Skin divers with both wealth and ambition can purchase underwater camera cases, particularly for minature cameras, and other underwater photographic equipment from the firms manufacturing diving equipment.

TIDES

61. Where does the water go when the tide is out? The water does not really go anywhere. The sea level rises and falls as the tidal wave passes through it, but the water itself does not move horizontally except where the sea is contained in narrow inlets or passages, for only in such places are tidal currents produced. The sea ebbs and flows on the shore simply because its own level goes up or down, like the water in a bathtub when you get in or out, except that here it is the moon and, to a lesser extent, the sun that make the disturbance. (See questions 63, 66, 76.)

62. Can you tell whether the tide is coming in or going out by looking at the shore? Sometimes you can, sometimes not. Along the spray-swept and often foggy Pacific Coast it is very difficult, and if you really want to know you had best sit down and watch some particular mark near the water's edge for a while. Likewise, on foggy and rainy days on the Atlantic Coast there is no other way unless you happen to know the time of high or low water. On sunny days, however, the weeds, rocks and sand dry out after the water leaves them, and by the time the tide returns, the seaweed higher up the shore may be hard and brittle and the rocks and sand warm and dry, so that the incoming tide rises over a parched shore landscape. An outgoing tide on the other hand has left the shore but recently and the seaweeds are soft and fresh, the rocks often slippery, and the sand cool and damp.

63. What makes the tide? The attractive force of the moon, primarily, and to a lesser extent that of the sun as well. As the earth rotates, each section of the watery envelope which is the sea comes to be in line with the moon's pull approximately twice a day. In consequence a pulse or wave is set going which we call the tide. It starts where the water is deepest and where there is least land to interfere, which is somewhere in the Southwest Pacific Ocean. From there the tidal wave travels west through the Indian Ocean and passes north through the Atlantic into the Arctic seas. It also moves from the center of origin east and north across the Pacific.

64. How often does the tide rise and fall in a day? Where the moon tide is the predominant tide, as it is throughout the North and South Atlantic Ocean, the tide rises and falls twice in every twenty-four hours. The Pacific Coast tides, however, are more complex, although for most of the Pacific Coast there are still two tides a day. In the Gulf of Mexico there is but one small tide a day.

65. Does high tide always come at the same hour? No. Each high tide is about fifty minutes later than the preceding morning tide and each afternoon tide about fifty minutes later than the afternoon before. If the moon remained in a fixed position relative to the earth, the tide would be at the same time each day. But the moon itself rotates about the earth in approximately twenty-nine days, and so is about fifty minutes later every day in reaching the same position. Consequently the high tides of one day are about fifty minutes later than the corresponding high tides of the day before. This is the typical state of affairs and is characteristic of both east and west sides of the North and South Atlantic and of the Indian Ocean. The Pacific Coast and the Gulf of Mexico, however, do not conform. (See questions 63, 69.)

66. Why are tides higher in some places than others? The tidal wave of the open Atlantic Ocean has a rise and fall of from three to four feet, depending on the phase of the moon, and that is the typical tidal range on most of the shoreline bordering the sea. The amplitude of the wave may, however, become dampened down in regions where the wave is obstructed by island barriers, so that in the Caribbean, for instance, the rise and fall of the tide decreases as you pass westward from the eastern chain of islands. Similarly, in the Mediterranean the Atlantic tidal wave fails to enter the Strait of Gibraltar and in consequence the Mediterranean is tideless. On the other hand, where the oceanic tidal wave passes into large blind alleys such as the Bay of Fundy and the Severn Channel, where the land converges on each side and the depth decreases, the wave piles up more and more as it passes along, until finally at the head of each of these places the tidal rise and fall is about fifty feet.

67. Why are tides higher at certain times of the month? (See questions 75, 76.)

68. How long an interval is there between high tide and low tide? Where tides come twice a day, as on the Atlantic Coast, the interval between high and low water is a little more than six hours.

69. Does high or low tide occur at the same time all along a coastline? No. The tide, being a wave which is traveling through the ocean, reaches one point along a shore earlier than it does another farther away in the direction of travel. Accordingly, on the Atlantic Coast the times for high and low water become later as you pass from Florida to Nova Scotia. The tide turns sooner, however, at the heads of creeks and other inlets than it does on the adjacent open coast.

70. How can you anticipate the time of the low tide? You can be observant and note the time of high tide, knowing that the next low tide will be roughly six hours later; or, having noted the time of the low tide on one occasion, you can add approximately fifty minutes for each day that has since passed. Or, for not more than fifty cents, you can obtain from the Superintendent of Documents at Washington, D.C., a government time-table for the year for either the whole of the Atlantic or the whole of the Pacific Coast.

71. What is meant by "slack water"? It is a period of an hour or so following high tide and low tide when the sea level is neither rising nor falling, more specifically known as "high-water slack" and "low-water slack." The period of low-water slack, before the actual turn of the tide, is the best time for exploring the low-level rocks and beaches.

72. What is the difference between the daily tides on the Pacific Coast and those of the Atlantic Coast? On the Atlantic Coast there is not very much difference between the two low tides or the two high tides of any particular day, but on the West Coast there may be a great deal. Thus each day on the Pacific Coast there is a high high-tide and a low high-tide, and a low low-tide and a high low-tide. Therefore, inasmuch as only a good low tide is an invitation to explore the shore, the seashore naturalist on the Pacific Coast has but one useful tide a day during which to go collecting or observing, instead of two. The reason for the unevenness of the Pacific tides takes us too far into the dynamics of wave motion in relation to the size of the ocean basin.

73. What are the best tides for collecting? Since the best plan for collecting or observing marine life on the shore is at the lowest level possible to reach, the best tides for this purpose are those that go out the farthest. For several days following the time of the full moon and the new moon the tides are lower than at other times.

74. When is the best time for exploring the low-tide level of the shore? During the hour or so before the tide turns, during the so-called low-water slack. At this time the water is usually comparatively calm and the level of the water more or less stationary. Once the tide has turned and has begun to rise, the water usually becomes more ruffled by wind and wave and shortly starts to rise fast.

75. When are the low tides at their lowest? At the so-called "spring tides," which come after every full moon and at the time of the new moon. At certain times of the year, however, even the spring tides are greater than at other times. On the Atlantic Coast the greatest tides come during the spring and autumn equinoxes. On the Pacific Coast they occur during midsummer and midwinter.

76. When does the tide go out farthest? The tides are greatest—that is, go out farther and come in higher—when the moon and sun are more or less in line with the earth and are exerting their pull in a mutually reinforcing rather than in an independent manner. This occurs at the time of the full moon, when the earth lies directly between the sun and the moon, and at the time of the dark of the moon, when the moon lies directly between the sun and the earth. At these times, two weeks apart, and for a few days afterwards, the tides are higher and lower than at other times. They are known as the "spring tides," whereas those that come at the times of the half moon are not so great and are known as the "neap tides."

77. What makes waves, breakers, undertows and currents? Waves are nearly always the result of the wind acting on the surface of the sea. The exceptions are the great tidal waves produced by the pull of the moon and the sun on the ocean as a whole, and the smaller but usually disastrous tidal waves caused by submarine earthquakes and volcanic eruptions. In the open sea the wind waves range from cat's-paws produced by small descending gusts of air to the tremendous waves produced by wind of gale strength acting steadily over stretches of sea hundreds of miles long, when fifty-mile-an-hour winds build up waves thirty-five feet high traveling at forty miles an hour. Such waves in the open ocean are known as oscillatory, that is, the wave is an undulation which passes through the water without causing the water itself to move forward. Only the disturbance travels horizontally, just as in the case of the surface waves produced by dropping a pebble into a pool. But, also as in the case of the pebble, the wave disturbance produced by the wind travels far beyond the storm area. Large waves seen along the coast on a calm, sunny day are evidence either of a distant storm or of one that has passed by sometime before.

When open-sea waves reach the shallow region close to shore, their nature changes. They become shorter and higher and the front of the wave becomes steeper, until a true forward movement appears and the waves actually move toward the shore. If the wave front is not too steep, you see a smooth so-called ground swell which ends with a crash against the coastal rocks or beaches. As a rule, however, the wave front becomes too steep and the waves become top-heavy, so that they topple over as they rush forward. So it is that breakers are produced, advancing on the shore in regular succession.

Surf breaking on rocks is always dangerous, since the water falling away after each broken wave has a mighty pull and few persons can hold on against its sucking weight. And in much the same way the sea breaking on a steep beach is always dangerous, whether it arrives as a series of breakers or as a quiet ground swell crashing only at the last moment. Here again gravity is the cause of the danger, for it produces an undertow. The forward movement carries the wave up the beach and the slope of the beach causes the water to run rapidly back again, even though another wave is coming in over it. The situation is particularly dangerous since it is not at all obvious to the eye. The steeper the beach or the longer the slope the greater the undertow will be. It may well be powerful at half tide and virtually absent at high and low water if only the middle range of the shore has definite slope.

Tidal currents are an additional danger in coastal waters. In narrow spaces the main tidal wave increases greatly in height and also causes an extensive to-and-fro movement of water known as tidal currents. In localities like the English Channel they run at about three miles an hour, but in very narrow passages like the channel between Scotland and the Orkney Islands they can run as strong as twelve miles an hour, currents that only high-powered ships can face. In some places strong currents run over rough and shallow bottoms, or two strong tidal currents may run into each other, to produce a noisy and turbulent state known as a tidal rip. In certain cases, because of the depth and the nature of the sea bottom, the current becomes a whirlpool capable of engulfing small ships, such as the Maelstrom off northern Norway, and Charybdis in the Straits of Messina separating Sicily from Italy.

SEAWATER

78. What gives seawater its distinctive taste? Seawater contains many salts, but the predominant one is sodium chloride or common table salt. One of the runners-up is magnesium sulphate, which is the same as Epsom salts. Between the two you get that highly distinctive flavor. The salts of the sea have come from the earth's crust, dissolved by water which has reached the sea by run-off from the land and by upsurging from the sea floor during submarine volcanic action, over an enormous period of time.

79. Why should you not drink seawater? Because seawater is so salt that it takes more water from you than you can get from it, so that you are left worse off than before. The blood of land and fresh-water animals, including our own, is much less salt than the sea, possibly because the sea has continued to get saltier since the time our ancestors originally left it, which was at least some hundred million years ago.

80. How salt is the sea? There are several answers and you can take your choice. Seawater is equivalent to a nearly two-thirds molar solution of sodium chloride. It contains, volume for volume, about three times as much salt as your blood. And at its full strength, undiluted by fresh water from any source, it will float a hen's egg, if the egg is fresh.

81. What makes the sea blue? The blue color of the deep sea is due to the scattering of light among the water molecules and is therefore comparable to the blue of the sky. The emptier the water the bluer it appears, the bluest water being the most destitute of life of any kind. The deep-blue ocean is in fact virtually a watery desert and you find richer life where the watery fields are greener. (See questions 82, 86.)

82. Why does the sea look like diluted green-pea soup in the summer in northern regions? The greenish or yellowish-green murky water you find along the more northern stretches of the Atlantic and Pacific Coasts during the summer months is the result of overabundance of diatoms. While they serve as food for many small creatures and even some of the larger fish such as the menhaden, they can become so thick that herring, for instance, will avoid swimming through the denser parts. (See questions 83–85.)

83. What are diatoms? Microscopic, single-celled plants which abound in the sea, each individual being closed in a delicate case of

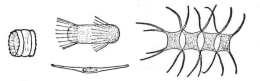

Diatoms (greatly magnified) (1/300″)

silica. They are called diatoms (di-atoms) because the glass-like protective shell is composed of two halves that fit one into the other, enclosing the living organism in a little box.

84. Of what importance are diatoms? Together with similarly small plant-like organisms called coccospheres and peridinians (or armored flagellates) the diatoms constitute the most important part of the

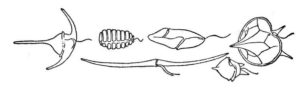

Armored flagellates (greatly magnified) (1/200")

drifting plant life, without which the seas and oceans would be almost empty of fish and other kinds of animal life. So far as sea life is concerned diatoms represent the primary source of all vitamins.

85. How can you see diatoms? A net made of the finest silk or muslin must be drawn through the water (anything coarser will allow them to pass through) and the catch will look like a greenish-brown scum. To see them the scum must be diluted with seawater so that the individuals separate from one another and then they need to be examined with a microscope.

86. Why is the sea a chalky green around the Florida Keys and the West Indies? Because so much lime salts dissolves out of the coral substance, and is also brought out of solution in sea water by bacteria, that the seas become supercharged with suspended particles of carbonates of calcium and magnesium.

87. What is "red tide"? "Red tide" or "red water" is produced by a reddish single-celled organism related to the Noctiluca (see question 99) which at times becomes so abundant along the coast of southern California and the Florida coasts that it colors the water red for miles. Such outbreaks of "red tide" cause the death of fish and other creatures which then are cast up along the beaches to decay. The stench of the water and on shore may well become intolerable and has been compared with the smell of the Nile when, in ancient times, that river "turned to blood."

88. Are any single-celled organisms large enough to see? Yes. Those mainly responsible for the star-phosphorescence of the sea,

Noctiluca, are about pinhead size and can be seen by the unaided eye (see question 99). Others such as the heavy-shelled foraminiferans of the sea floor may be even larger, though they are difficult to sort out from the surrounding material. Most single-celled organisms, whether animal or plant, require the use of a lens or microscope in order to be recognized. An obvious exception is the green mermaid's cup of the Florida Coast. (See question 115.)

89. What is "red feed"? "Red feed" is the abundant, reddish copepod crustacean known as Calanus, which is one of the most important food organisms of such fish as the common herring. The crustaceans feed directly upon the microscopic plant life of the seas and so form

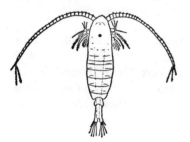

"Red feed" (the copepod Calanus)
(1/10")

an essential link between the plant and the rest of the marine-animal world, for they correspond to the grazing animals of the land. They are readily caught by towing a fine silk net through the water at times of their abundance, but must be examined in a jar of water to be appreciated.

PLANKTON

90. What is plankton? All the drifting life of the oceans, from unicellular organisms to large jellyfish, constitutes the plankton, in contrast to powerful swimmers such as fish, squid and whales, which together constitute the nekton.

91. How can plankton best be gathered? By towing a fine-meshed conical net of silk or nylon through the water, with a collecting container or jar at the end of it. The finer the net the smaller the organisms it will retain, but the less it filters. A coarser net retains only the larger kinds, mostly visible to the naked eye, but filters much more water and so catches much more of this kind of plankton.

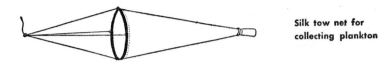

Silk tow net for
collecting plankton

92. Of what use is plankton? Apart from constituting the basic pasturage of the sea so far as the larger forms of marine life are concerned, it is under consideration as an important source of food for the expanding human population. In certain regions of the seas it can be strained out in fairly large quantities and, when the water is squeezed out, yields a greenish mush with a shrimpy taste. As a solution for mankind's swelling food problem the use of plankton is far from being as promising as many people appear to believe.

93. Is plankton found only near the surface of the water? No. The greatest abundance of plant plankton occurs between 100 and 200 feet from the surface, where light intensities are generally most suitable for its growth. Below 600 feet there is little light and plant cells can no longer grow. The animal plankton which feeds directly upon the living cells are consequently confined to much the same depths and so are those that feed upon them in turn. There is however a constant rain of dead and waste matter falling into the depths below and many creatures rely upon this manna from above for food. Small crustaceans have their greatest abundance below the 600-foot depth.

94. Why is plankton more abundant in some regions than in others? In certain regions, such as most of the coastline of North and South America and locally in other places, cold water rich in fertilizer salts wells up from the depths and maintains a good supply at all times, so that the plant plankton which supports the rest is always abundantly present. In other regions where there may be plenty of light but no upswelling or other means of replacement, the salts are constantly on the verge of exhaustion, so that clear-blue deserts of water are the result. Apart from upswellings of phosphate-rich water, areas along coastlines where drainage from the land is considerable are also comparatively rich in these essential salts and in many regions the coastal waters are particularly rich in plankton as a result.

95. Why is plankton more abundant at some periods than at others?
Plankton as a whole depends on the growth and multiplication of its plant components, and those in turn are dependent upon the presence of certain mineral salts, which are in somewhat short supply, and upon relatively long periods of daylight. The salts are principally the phosphates and nitrates—the same basic fertilizer salts of the land—and they are present in the upper sunlit layers after the winter storms have mixed the waters and brought salts up from below. Through spring and summer months the plant plankton multiplies abundantly and so does all that lives upon it directly or indirectly. By fall, when days are shortening, the salts have for the most part been used up and both the plant and animal plankton diminishes accordingly.

96. What feeds on plankton? The smaller components of plankton are fed upon by the larger, generally in a step-by-step fashion, so that minute animals such as the crustacean copepods and larval stages of many kinds feed upon the single-celled plants and animals. Others in turn, such as arrow worms (*Sagitta*) and young fish, feed upon them, and so on up the ladder of size, the whole forming so-called "food-chains." At the top of the scale the largest of all creatures cut out the middlemen—the giant whalebone whales, up to eighty feet long, feed mainly on small planktonic shrimp, and so do the largest of the sharks, the forty-foot whale shark, one of the most harmless creatures in the sea. (See questions 82, 89.)

97. What is detritus? Detritus is the name given to the rain of fine particles constantly sinking to the sea floor and consisting of waste or the disintegrating material of the animals and plants of the plankton (see question 93). Many kinds and astonishing numbers of animals that live on the sea floor depend upon the detritus for food and have various mechanisms for filtering it from the water. (See questions 375, 650, 712, 756.)

PHOSPHORESCENCE

98. What is phosphorescence? Sea phosphorescence is the result of animal luminescence. Many marine animals ranging from deep-sea squid and fish to microscopic single-celled organisms give off a greenish-white light under certain circumstances. In the case of the more complex creatures the production and use of the light is under control. In others of a simpler kind, whether large jellyfish or microscopic cells, the creature needs to be mechanically disturbed or irritated in order to flash. (See questions 46, 88, 767, 768, 870, 926.)

99. What is the "star phosphorescence" of the sea? The numerous pin-point flashes of light which illumine the water along the shores on dark summer nights when any moving body such as fish, boat or bather passes through it are produced by single-celled organisms of the group known as the armored flagellates, which are partly plant-like, partly animal by nature. The organisms are large for their kind, being about one twenty-fifth of an inch in diameter, and are known

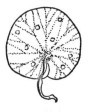

Noctiluca, the star-phosphorescent organism (1/50")

as Noctiluca, which means "night light." Frequently in inshore waters the Noctiluca may be so numerous that any object passing through the water becomes sharply outlined with light as contacts are made. On one occasion a quart of seawater dipped from the surface of the Gulf of California was found to contain about three million Noctiluca, which is a lot when you consider that an individual is large enough to be seen by the naked eye. (See question 88.)

100. How is it produced? Whether luminescence is produced by fish, worms, bacteria, single-celled animals or by the fireflies, the chemical basis seems to be the same. Virtually no heat is given off, and the chemical process is not to be compared with combustion but with glandular activity in which an enzyme or ferment acts on another substance to produce a third substance plus energy, which in this case is in the form of light.

101. Why are some animals phosphorescent? Certain deep-water fish, shrimps and squid have luminescent organs which they employ either for maintaining communication among themselves or for attracting prey, or both. Most luminescent creatures, however, appear to shine more by accident than design, as though the luminescence or phosphorescence is an inadvertent by-product of nervous stimulation, for in most cases no useful purpose appears to exist. In all likelihood the production of animal light is simply an accompaniment of certain chemical changes associated, for instance, with the sudden production of slime by glands, but in certain cases the production of light has been brought under definite control and put to some use.

102. When can you best see phosphorescence? On dark moonless nights during spring and summer months, from about one hour after sunset onward.

SEAWEEDS

103. How is it that seaweeds on the shore are usually distributed in definite zones? The spores of all weeds settle on the shore from the high-tide mark down to below low-tide level, but in each case only those survive which find conditions well suited to them. Thus the seaweed known as channeled wrack (*Pelvetia*) occupies the uppermost zone or band, immediately followed by a narrow belt of flat wrack (*Fucus spiralis*). Below this again comes the broad expanse of the intertidal rocks draped with knotted wrack (*Ascophyllum*) and bladder wrack (*Fucus vesiculosus*). Then at the water's edge the short-cropped belt of Irish moss takes over, leaving the deeper water beyond to the kelps. But the seed of all is sown everywhere, only to be weeded out by varying degrees of intertidal exposure where conditions are not suited to any particular kind.

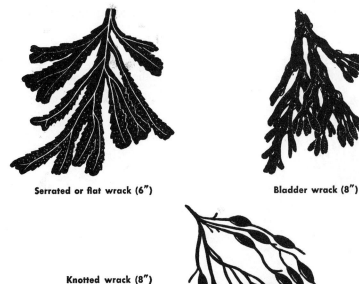

Serrated or flat wrack (6″)   Bladder wrack (8″)

Knotted wrack (8″)

104. Do seaweeds have roots? Not in the sense that land plants have roots. Seaweeds with few exceptions are anchored to rocks and other submerged solid surfaces and in many cases their holdfast processes have the appearance of roots. Such processes are merely anchors and do not in any way supply the weed with food materials as in the case of roots.

105. How do seaweeds stay upright in the water? Most of them consist of material which is a little lighter than seawater and in most cases seaweed broken away from its attachment floats and drifts at the surface for this reason. Certain kinds which may be larger and denser than others have built-in floats or air bladders which make them especially buoyant.

106. How do seaweeds grow? Nutritively speaking, they grow like other plants, utilizing mineral salts and the energy of sunlight, but they obtain their salts from the surrounding seawater and not from the ground by means of roots.

107. If a blade of kelp is broken off from its stalk, will a new one grow? If the blade is torn off above or at the junction with the stalk,

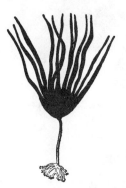

Palmated kelp (6')

a new blade will grow, for the base of the blade is the region where growth normally occurs. For the same reason most seaweeds will grow again when torn off the rocks, so long as the stumps are left attached.

108. How large can seaweeds grow? Some of the kelps, such as those growing offshore along the California coast, can grow to a length exceeding 100 feet, forming great submarine seaweed forests which shelter all kinds of animal life including the giant tunas. Similar forests are found elsewhere, for example in the Strait of Magellan, wherever the sea is cold and contains the right amounts of certain salts.

109. How do seaweeds reproduce? At certain times of the year, differing according to the kind of weed, the tips of some of the blades or fronds become swollen. This is particularly evident in the weeds occupying the middle and upper zones of the shore. The swollen parts contain small cavities in which male and female cells are produced, one or the other according to the sex of the individual plant. When they are ripe, after the tide goes out, masses of jelly containing the reproductive cells ooze out onto the surface of the weed. When the tide comes in again, the eggs and sperm are released and come together in pairs. The fertilized spores then float away and finally settle somewhere, to become attached and grow if on a rock, to die if on mud or sand.

110. What are the colors of seaweed? They are grouped into the blue-greens, the greens, the reds and the browns. The so-called blue-green algae, however, are not always blue-green, and one microscopic kind is responsible for the red color of the Red Sea. The group is peculiar both in its metabolism and in its sexless manner of reproduction, but it does not form a significant feature of the seashore. The

Peacock tail seaweed (1–2")

greens, which may be a vivid light green as in the sea lettuce (*Ulva*) and sea moss (*Bryopsis*) and sea hair (*Enteromorpha*), are generally found on the shores wherever fresh water seeps over the rocks, or a rich dark green like the sponge seaweed (*Codium*), add color to the rocks and pools but do not make major contributions in a quantitative sense. The reds and browns comprise the bulk of the seaweeds.

111. Why are most seaweeds on the shore brownish in color? They contain a greenish-brown pigment which absorbs light like green chlorophyll and has much in common with it, although it is effective in water of depths down to seventy-five feet. The brown algae as a group however are generally thick and tough compared with other kinds and contain large amounts of a gelatinous substance called algin. This substance dries out or takes up water readily without deteriorating, and the weeds that possess it can withstand the ebb and flow of the tide on the shore better than other kinds. Kelps and rockweeds are all brown algae, and so is the drifting Sargassum or gulfweed.

112. Why are red seaweeds red? The small, usually delicate, red seaweeds of the lower rock pools and the rocks below low tide are red because they possess a red pigment which absorbs light for energy purposes in much the same way as chlorophyll, although more of the blue and violet light than chlorophyll does. Since blue and violet light penetrates deeper in the sea than other colors, the red algae can live and grow at greater depths than the other kinds, down to about 200 feet. They are not equipped to survive long out of water, so that those to be seen along the shore are either in rock pools or on rocks at the low-tide level. Irish moss, dulse and the limy corallines are common kinds.

113. What are the pink, limy growths seen lining low-level rock pools and other places? They are the corallines, which are seaweeds that have a limy skeleton like that of many animals. Some coralline seaweeds have the typical branching appearance (*Corallina*) while

Coralline seaweed (1″)

others form extensive encrusting sheets (*Melobesia*) and look like pink lichens. Sea urchins, which need a lot of lime in their diet to maintain their own limy skeletons, feed on them to a great extent. (See question 202.)

114. What is the greenish-black scum that makes rock ledges so slippery after rain or just after the tide has left them? It is as much a seaweed as any other, of the group known as the green algae. It dries out in the sun so that it looks like lichen, but when wetted by either salt or fresh water it swells and makes the rocks treacherously slippery.

115. What kind of plant is the little mermaid's cap? The mermaid's cap (*Acetabularia*) found in pools of shallow water along the Florida

Mermaid's cap (1–2″)

Coast is a green alga which not only looks like a small greenish-white mushroom but consists of a single enormous cell instead of thousands of microscopic ones.

116. Why are rock weeds and many kelps divided into long narrow strands like the fronds of a fern? Because otherwise the force of the waves would tear them to pieces. As it is, the water streams between the strands without doing any harm, the weeds in turn streaming this way and that as the currents swirl about.

117. Are seaweeds harmed by drying out when the tide goes out? Those that properly belong on the shore rocks, as distinct from the more delicate inhabitants of pools, can dry out hard and almost brittle with impunity. They lose water just as gelatin does, and swell up again as soon as they are re-immersed. (See question 62.)

118. Why are some rocks covered with seaweed and some not? There are several reasons, any or all of which may be responsible in a particular case. Seaweeds first settle on rocks as microscopic sporelings. They cannot settle where waves and currents are so strong that they have no chance to become attached. Where they do attach they are likely to be eaten, for the young of many common snails and the adults as well feed extensively upon them and often keep rocks clean of any obvious growth. Lastly, each kind of weed has its own requirements, and what one can tolerate in the way of exposure or muddy water, another cannot. (See questions 103, 109.)

119. Can you keep rocks along the shore clear of weed? The weed can be cut off but fragments left behind will grow again and in any case the rocks will be resettled by seaweed spores so long as the conditions remain the same.

120. Are all seaweeds originally attached to rocks or some other · solid surface? Almost all are so attached, and most of the weed tossed along the beaches at the high water mark has been torn loose by rough water. One kind, however, the gulfweed or Sargassum, the most famous of all the brown algae, has been adjusted to living and growing as large drifting masses in the open ocean. It may be tossed ashore by the tide but is also commonly seen drifting in the water off the Atlantic Coast of Florida and northward to Cape Cod.

121. How does gulfweed stay afloat? Small air bladders which look like berries attached by stalks to the stem of the weed make the

Gulfweed or sargassum (8″)

Sargassum very buoyant. Huge areas of the weed drift slowly in an enormous eddy in the Atlantic Ocean to make the so-called Sargasso Sea.

122. What animals are likely to be found in a mass of floating gulfweed? A variety of creatures, all of the same yellow-brown colors as the weed and therefore hard to see. They include a small fishing-frog fish, a crab and a flatworm, none of which are to be found elsewhere. Various hydroids also attach to the weed, while seahorses occasionally take an overlong ride hanging on to the weed with their tails, to end up in regions much too far to the north. Lastly, flying fish attach their eggs to the drifting weed, and the newly hatched young use the weed as nurseries to hide in.

123. What use do sea creatures make of seaweeds? Many use the curtains and forests of seaweeds simply as shelter or hiding places. Lift up the rock weed off the lower shore rocks and you may find many of them. Others such as sea urchins, menatees, and certain sea snails, feed on weeds of various kinds. While many small creatures, such as the spiral-tube worms and sea mats and hydroids, find permanent settling places on the broad surfaces of the blades. Each kind of weed bed, whether kelp, rockweed or eelgrass, harbors its own community of animal life.

124. Is any seaweed used as food for human consumption? The purple laver and dulse, which are very thin red seaweeds, like a purple sea-lettuce, make excellent soup and are often eaten raw, particularly in the Canadian Maritime Provinces. Indirectly many seaweeds, particularly the kelps and sea wracks, help feed humans inasmuch as they serve as soil fertilizer on coastal farms in many parts of the world.

125. What is kelp used for? The giant kelp (*Macrocystis*) of the California coast, which is collected in great quantities, is used as a

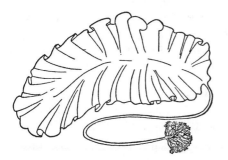

Laminarian kelp (10')

source of acetone, potash and iodine, while a substance called algin is extracted from the broad-leaved kelp (*Laminaria*) of the east coast and used in making ice cream, cake icings, and dental impressions.

126. What is Irish moss used for? It is collected chiefly from the coasts of Massachusetts, Maine and Nova Scotia and a substance

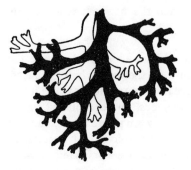

Irish moss (1–2″)

called carrageenin is extracted for use in the making of candy, jellies and salad dressing. Other substances are also extracted and used in various industrial processes.

127. Where is Irish moss found? Irish moss (*Chondrus*) is found a little above and below the water's edge at low tide along the more northern parts of the rocky coasts. The weed is a yellowish green or purplish, with iridescent blue lights, and has a generally cropped appearance.

128. Can you come to any harm swimming among seaweed? No more than elsewhere. The weed is clean and generally has a calming effect on waves. On the Northwest Pacific Coast, for instance, sea otters wrap pieces of kelp blade around their bodies and take naps whenever they feel so inclined—that is, the few sea otters that we have left alive. (See question 1043.)

129. Do seaweeds ever flower? True seaweeds do not. They have sex but not flowers or anything comparable to them (see question 109). Several flowering plants, however, have invaded the sea in sheltered shallow regions. These are principally eelgrass and turtle grass.

130. What is eelgrass? Eelgrass (*Zostera*) is the seed plant which has invaded the estuaries and sheltered inlets and has become adapted to living in salt water. It has true roots and produces inconspicuous flowers and seeds. Eelgrass forms large beds of long grass-like blades whose roots mat together below the mud. Many animals find their place as part of an eelgrass community, though less so than formerly since the eelgrass disease swept the North Atlantic. (See questions 132, 133.)

131. What animals are found among eelgrass? A number of small snails, certain limpets, scallops, pipefish and perhaps above all the stalked jellyfish (*Haliclystis*) which attaches to the blades. (See question 38.)

132. What destroyed most of the eelgrass beds of the Atlantic Coast? A water-borne disease, which was either a virus or a fungus, swept up the Atlantic Coast of America at about the time the great depression swept across the continent, and after about a year succeeded in crossing the Atlantic Ocean to infect and destroy the eelgrass beds of Northern Europe. Here and there the beds show signs of recovery, but most eelgrass beds now seen are of a smaller variety which remained immune.

133. What is turtle grass? It is a true grass (*Thalassia*) closely related to eelgrass but found in tropical waters including southern Florida. Like eelgrass beds, turtle grass shelters its own special community of animal life. (See question 37.)

134. Why is turtle grass so named? Because sea turtles hunt in the beds of turtle grass for the mollusks and other creatures which live there. (See question 1017.)

JETSAM

135. Why do you usually see dried seaweed in several distinct lines along the upper part of a beach? Each line represents the weed left behind at the level of high water. If there is only one line of weed, that line represents the last high tide and also indicates that the tide was higher than the preceding one. If there is more than one line of weed, the lower line indicates the height of the last high tide and also that the last tide did not reach so far up the beach as the one before. (See questions 75, 77.)

136. What are the numerous jumping flea-like creatures that you see when you disturb the seaweed rotting at high tide? They are the crustaceans known as sand hoppers, which feed on the weed and burrow in the sand. (See question 642.)

137. What are the greenish-grey collars made of sandy jelly which look like clergymen's collars? They are the egg masses of the moon snail (see questions 334, 339) which have become dislodged from their position in the sand below low water. The eggs are still present, though probably dry, embedded in the material.

138. What are the empty black cases, like coin purses with a prong at each corner, seen among the seaweed on the beach at high tide? They are the egg cases of the common skate detached from their anchorage after the young skate has hatched. (See question 973.)

139. What are the strings of parchment-like discs, about two feet long, found with the high-tide debris or lying on the sand flats lower down? They are the empty egg cases of the large whelks. Each disc contained many eggs which finally emerged as tiny whelks. The whole string is a single spawning by one individual. (See question 273.)

140. Are the shells of crabs found along the beach the shells of dead crabs? In most cases, no. The majority of crab shells are shed skins or molts which have been cast off during growth and have been carried up the beach by the tide. (See questions 528, 535.)

141. How can you tell whether a crab shell is a molt or a part of a dead crab? It depends on how much of the shell is there. If it is not too badly damaged or incomplete you can tell by lifting the top part of the shell off and examining the underlying part. If it is a molt you will see the cast of the gills on each side and even the lining of the stomach complete with its stomach teeth. If it is a dead crab, there will be the remains of flesh as well, and moreover there will be no slit at the back of the shell where you might have lifted the top up.

142. What creatures are likely to be found attached to or inside driftwood? On the outside you may find acorn barnacles or, if you are lucky, the gooseneck barnacles (see question 648). Sometimes the wood contains the chalky galleries of the shipworm (see question 432) or the termite-like burrows of the gribble. (See question 639.)

143. Are the heavy horseshoe crabs found tossed up in large numbers at the high-tide level on certain beaches dead animals or what? They are the cast shells which have become partly filled with sand, etc. (See question 679.)

144. How does sand get into the cast of a horseshoe crab? By the same passage by which the animal got out. If you examine the cast shell carefully you will see that the edge along the whole length of the great curve is slit, although not gaping. This is where the horseshoe crab crawled out of its old shell and where the sand has sifted in.

145. Why do you find so many casts of horseshoe crabs roughly the same size on some beaches at certain times? Because every year a new generation of horseshoe crabs starts life at about the same time, and in any particular region grow at much the same pace. At definite intervals they cast their shells, and those of a given size and age will most likely shed their shells at about the same time. Consequently there are likely to be many shed shells or casts of more or less uniform size thrown up by the tides on the beach during a particular period, or else hardly any.

146. Why are mollusk shells found along the beach at high tide usually so white? With the exception of the blue or mussel shells most shell pigments bleach out under the influence of sunlight. Only shells recently cast up retain their original coloring.

147. What are mollusk shells made of? Mollusk shells are much more complex than at first appears. The shell consists of three layers. The outer layer is very thin and horny and protects the hard lime of the shell from being dissolved by the water. This horny layer may be rubbed off. It is seen, for instance, as long projections in the case of the bearded mussel, and as the hinge ligament of all bivalves. The next or middle layer is the thickest layer and consists of crystals of calcium carbonate arranged perpendicularly with the surface of the shell. The inner layer is the pearly layer and consists of crystals of calcium carbonate laid down parallel with the surface of the shell.

148. How are shells made? The horny and middle layers are produced only by the edge of the mantle tissue and show concentric growth rings since the material is not laid down continuously. The inner pearly layer, however, is laid down by the whole of the inner surface of the mantle and has a lustrous appearance.

149. How can you tell how old a shell is? You cannot tell how old a shell is, but you can tell how old the animal and its shell were when the animal died, at least in the case of many bivalves and some snails. It depends on how great are the seasonal changes in temperature between summer and winter. If they are large, then a bivalve lays down its shell at sufficiently different rates in winter and summer to produce annual rings like those you see in a cross-section of a tree. Most clams, scallops, etc., have shells showing many concentric rings placed close together, but these rings are grouped into recognizable bands, each band corresponding to a year's growth. Annual growth rings can also be seen in the shell of many snails, particularly in those with smooth surface and obvious spiral as in the case of the moon snail. The operculum of marine snails often shows lines representing annual growths.

150. Of what use are the various ridges, knobs and projections seen on many shells? For the most part they have no useful purpose but are either mainly ornamental—that is, are simply unimportant excesses of shell material—or represent folds in the living mantle which produced them. What the mantle folds signify is another matter. The fan-like ribs of a scallop shell, for instance, are the result of the waved or frilled structure of the edge of the mantle which lays down the shell. Whatever shape the mantle tissue takes, the shell takes too. How shells are made is something worthy of much more study than the problem has so far received.

151. Why are some shells naturally polished on the outside while most are not? Shells are polished wherever the mantle tissue of the mollusk is in contact with it, which is why the insides of all shells are so smooth. In some cases, such as the cowries and cone shells, the animal extends its mantle fold up the outside of the shell, so that the shell is almost covered when the mollusk is fully expanded. This keeps the shell surface in a glossy condition. (See question 289.)

152. What are the neat, round holes seen in so many clam and mussel shells? They are evidence of murder. Every shell of a bivalve or snail showing a neat, circular hole about one eighth of an inch across or smaller has been drilled by a carnivorous snail such as the moon snail, whelk, or oyster-drill. The shell was drilled while the occupant was alive, after which the victim was consumed. (See questions 337, 275, 260.)

153. What are the markings seen on the insides of clam or scallop shells? In fresh shells the inside layer usually has dark indented patches and a long dark line running parallel to the outer edge. In shells that have lain on the beach for a long time, the same indentations can be traced but the color has bleached out so that the markings are no longer obvious. In any case the markings or indentations are the places where the muscles were attached which held the two shells together and where the edge of the mantle was attached to the shell. At these places the inner lining or coating of shell material was interrupted and so marks or depressions were made.

154. What is the long tubular projection extending from the front edge of a whelk shell? In life it covers and protects the greater part of the long tubular siphon through which the animal draws in its respiratory current of water.

155. What is the use of the tooth or short spine seen on the outer tip of certain shells? Those snails which have a tooth on the outer lip of the shell live on barnacles and small mussels, etc., and use the tooth as a wedge to open the two-piece door of the barnacle or two shells of the bivalve. (See question 264.)

II. SHORE ANIMALS

ECHINODERMS

156. Where are starfish likely to be found? In general they are most likely to be found where their preferred food is abundant, but since few can stand exposure out of water for any length of time, they are to be found either in low-tide rock pools or just below the low-tide level itself—or, in the case of smaller species, on the underside of boulders low down in the tidal zone or on rocks beneath curtains of seaweed. They are most likely to be found where mussels are plentiful at or below the level of low tide.

157. Are starfish found anywhere except in the sea? No. All starfish, like all other echinoderm animals, are exclusively marine. Echinoderm is the group name and means spiny-skinned.

158. What are the limy structures embedded in the skin of a starfish? They are calcareous plates which give a certain degree of rigidity to the soft flesh, although not prohibiting bending and twisting of the body and rays. From the plates project calcareous spines, some of them movable. Together they comprise the skeleton, such as it is. Flexibility results from the fact that the plates are not united into a single shell but are embedded in soft flesh, and are distinct from each other.

159. How is the upper surface kept clean and free from parasites and other small creatures? The spines give some protection, but scattered over the surface and particularly around the gills are numerous small pincers known as pedicellarias which nip anything which comes in contact with them.

160. What is the small, hard, round limy disc on the upper side of the starfish's body near the junction of two rays? This is the sieve plate or madreporite, which is a finely perforated plate which allows water to pass into the interior but nothing else, not even bacteria. (See question 168.)

161. Do all starfish have rays? All starfish have rays, but the rays vary greatly in proportion to the central portion of the body. In the majority the rays are relatively long and the central part with which they merge seems only large enough to hold them together. In others the central region may be so large in relation to the rays that the whole star virtually becomes a pentagon.

162. How many rays do starfish have? Most species have five rays or arms, since the five-rayed pattern is basic to the large group to

Starfish shapes

which they belong. One kind (*Culcita tetragona,* from Europe) normally has four, but others may have six, seven, eight, ten, twelve or even as many as forty. (See question 192.)

163. Can starfish see? Not in the usual sense, but there is a minute reddish pigment spot at the end of each ray which is sensitive to light, so that the animal senses whether it is moving into dark crevices or into places more strongly illuminated.

164. Does a starfish have a brain? Not in the usual sense. It has a nerve ring around the mouth, with a nerve leading from it along each ray.

165. When a starfish moves, which ray leads the way? Whichever ray is pointing closest to the direction in which the animal is stimulated to move. When traveling, one ray points in the direction of travel and the others lie symmetrically two and two on each side. If a new direction is called for, the animal slows down and the ray pointing more or less in the new direction becomes the leader.

166. What are the waving tubes on the undersides of starfish rays?
The tubes are part of the locomotory mechanism and serve both for
movement and in the process of feeding.

167. How do starfish hold onto wave-swept rocks? They hold fast
by means of the suckers at the ends of their very numerous tube feet.
At the same time, starfish are so shaped and flattened out that the
pressure of the wave itself helps to press the animals against the rock,
rather than to dislodge them.

168. How does a starfish move? Locomotion in the starfish, as in
sea urchins too, is by means of a kind of hydraulic-pressure mechanism unique in the animal kingdom and known as the water-vascular
system. Water enters this system through minute pores in the sieve
plate located on the upper surface, and is drawn down through a tube
which joins a ring canal circling the mouth. From this ring arise as
many radial canals as the starfish has rays, usually five. Each radial

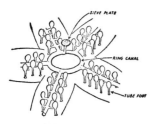

Hydraulic power system of a starfish

canal is connected with many pairs of tube feet which project from
under the surface of the ray. Each so-called tube foot has an ampulla
or hollow contractile bulb at its inner end. When the bulb contracts
the tube lengthens; when it relaxes the tube is drawn back. In most
cases the tube feet end in sucking discs that enable them to adhere
firmly to any object against which they are pressed. They are in fact
so efficient that many of the discs often break loose from the tube
feet when they are pulled off a rock. The action of all the tube feet is
co-ordinated so that they work together and draw the animal along.
Movement is effected by extending the tube feet ahead, attaching the
suckers to an object on the ground and then shortening the tube feet
to exert a forward pull. The tube feet are then relaxed and again

extended ahead for a new attachment. In some forms, such as species of Astropecten and Luidia, that crawl on sand or mud where there is nothing to attach to, the tubes have no suckers but are inserted into the sand and are used as a sort of movable traction device by which to move the animal along. (See question 862.)

169. Why does a starfish out of water become limp? Because water continually leaks from the ends of the tube feet and elsewhere and can no longer be drawn in through the sieve plate on the upper surface.

170. How does a starfish turn over? A starfish rights itself by bending its stiff arms and pulling with its tube feet.

171. If part of a ray is broken off, will the part remaining attached to the body become whole again? Yes. A new end grows slowly from the old stump, no matter where the break occurred, but the piece broken off cannot grow a new base and does not survive. (See questions 172, 174.)

172. What does it mean when some rays are small and some are large? The starfish has been injured and rays in the positions of the smaller ones were lost. The small rays are new rays in the process of being regenerated.

173. If a starfish is torn apart into single rays, can each become a new star? Yes, under certain circumstances. If the ray retains part of the central body region, the missing rays and the rest of the body region will grow out and reconstitute a whole starfish. A part of a ray without any of the central body will not do so.

174. Do any starfish break up naturally as a means of propagation? Yes. Several species exploit their power of regeneration for the purpose of propagation. Two of these are common on the Pacific Coast. One is an active, slimy, dark-colored starfish (*Astrometis*) found from Santa Barbara to Lower California, which divides into five equal parts as each of the five rays pulls away in its own direction. Unless eaten or otherwise accidentally destroyed, at least some of the rays, those retaining part of the central region of the body, regenerate the missing structures and become whole starfish. Better known for this activity, however, is another Southern Californian starfish which is remarkable in several ways. Not only does it pull itself apart and each ray with part of the center grow into a new ray, but it is the only genus known where parts of a ray can also reconstitute the whole animal. Ordinarily one ray breaks away at a time, and the usual method is for the main portion of the starfish to remain fixed and passive, and for one ray to walk slowly away at right angles to the body, changing position and twisting and doing all the work necessary for breakage. What the stimulus may be is not known. Moreover, if the tip of a ray is cut off, even if no more than a half inch, not only does a new tip grow in its place, but the cut-off tip grows a new body and five or six small new rays from its cut end. Sometimes a much longer piece will, after regenerating new arms for about a year and a half, throw off the new set. The new set then grows a new ray of its own, making five, while the original piece begins to regenerate a new starfish. The process frequently gets a little out of hand and specimens are commonly found with four, six, seven, or eight rays instead of the standard five.

175. How does a starfish breathe? Delicate small finger-like skin gills, which may be seen with the aid of a pocket lens, project from the upper surface of the body through spaces between the calcareous plates.

176. Where is the mouth of a starfish? The mouth is situated at the center of the underneath surface. The mouth opening is bounded by limy plates which form part of the skeleton.

177. What do starfish eat? In general, a starfish will eat any animal it can catch and envelop, though inasmuch as starfish move but slowly it feeds for the most part on those that cannot move away, such as oysters, mussels, clams and other bivalve mollusks. It also feeds on marine snails as opportunity serves, on barnacles, occasionally on carelessly moving crabs, and on moribund fish. Each kind, however, has its own particular preferences. Some limit themselves to a single article of diet, or little more. A small English species (*Asterina*), feeds on sponges and ascidians. On the Pacific Coast the sun star (*Pycnopodia*) will eat practically anything, but in rocky regions feeds mainly on sea urchins, while on the Atlantic Coast the common shore starfish (*Asterias*) feeds almost exclusively on rock barnacles, mussels and small whelks. (See question 182.)

178. Does the starfish have teeth? No. It is therefore unable to break up or swallow its prey in the usual way.

179. How does a starfish eat? When a starfish has succeeded in opening a bivalve it protrudes its stomach, turning it inside out in doing so, and thrusting the stomach lining between the shells of the mollusk. The starfish's digestive juices then pour into the soft parts of the victim. An oyster or mussel or clam, as well as barnacles (which

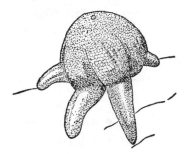

Starfish (6–12") in feeding position

are crustaceans), are thus digested within their own shells, while the digested flesh is absorbed by the everted starfish stomach until the inside of the shell is completely empty. The exceptions to this are the sand-dwelling starfish, (the species *Astropecten*), which swallow their food whole, the food consisting of small living snails living in the sand. It has a relatively wide mouth in proportion to the size.

180. How large do starfish grow? It depends on the species. On the Atlantic Coast, the northern star (*Asterias*) reaches seventeen inches across, while farther south the giant starfish (*Oreaster*) may be twenty inches. On the Pacific Coast the common starfish, (*Pisaster*) may attain to fourteen inches.

181. Are starfish harmful? Starfish are harmless as far as people are concerned, but certain species move about on the floor of shallow seas in hordes and can destroy an oyster bed overnight.

182. How does a starfish open the tightly closed shells of an oyster? According to R. W. Miner, "The starfish simply mounts on the oyster, humping up the central part of its body and spreading its five arms so that they surround the two valves of the mollusk, applying hundreds of tube-feet, some on one valve and some on the other, and bracing the tips of its arms on the surrounding objects. It then contracts the muscles of the middle part of its arms, exerting a steady pull on the two shells of the oyster in opposite directions. The single muscle by which the oyster keeps its shell together gives way sooner or later, yielding to the steady, tireless pull of the multitudinous tube-feet, and the oyster-valves begin to part." E. K. Ricketts says, "The starfish's humping-up process enables it to exert the necessary pull in opposite directions. It has been demonstrated that if a specimen is lightly compressed between two glass plates so that it cannot assume the humped-up position it cannot avail itself of food."

How much of a pull a starfish can thus exert, however, is not known and there may be more than a tug-of-war involved. A paralyzing chemical action is at least suspected, for after a starfish has attacked an expanded moon snail for only a few minutes and is then removed, even though no physical damage can be seen, the moon snail remains expanded and motionless thereafter until it begins to decay. In the case of bivalves it is likely that starfish stomach juices seep in between the shells and slowly cause the molluscan muscles to relax.

183. Since starfish are a pest on shellfish beds do fishermen tear apart those they catch in order to destroy them? No. This used to be a common practice, but now it is well known that tearing starfish apart and throwing them back in the sea merely increases their numbers.

184. Have starfish any economic value? No, except for insignificant ornamental purposes.

185. What enemies do starfish have? Starfish have very few enemies, the chief ones being the seagulls which search along rocky shores at low tide.

186. Do starfish stay in one place all their lives? Different kinds move about to varying extents. Small species tend to remain within very restricted territory among rock weeds. Larger species are inclined to move or migrate in groups from place to place as they exhaust one feeding ground after another, and also according to changes in the temperature of the water.

187. How do starfish propagate? With very few exceptions starfish have separate sexes and in most cases the eggs and sperm are spawned into the water, the fertilized eggs developing into ciliated larvae which are free-swimming for a period of from two weeks to two months. The larvae then undergo metamorphic changes and settle on the sea floor as miniatures of their parents. Only a small percentage survive to this stage, since the vast majority become either lost or eaten or settle in the wrong places. Some starfish, like the sea bat (*Patiria miniata*) which ranges along the whole Pacific Coast from Sitka to La Paz, can breed at any time of the year. Others breed only at certain periods of the year, depending on the species. A few starfish, particularly those that are small when mature, brood their eggs and young. The three species of the six-rayed starfish of the Pacific Coast, and the smooth blood sea star, of both the Pacific and Atlantic Coasts, carry their embryos on the undersurface near the mouth and may even hump up over them like a tent to form a temporary brood pouch. The larvae of these never swim to the surface but commence their free existence by gliding on the bottom. A common star of European shores (*Asterina*) also broods its young, but in addition undergoes a change of sex. All young forms are male, but as they grow larger they become females and produce eggs only. (See question 174.)

188. Can you tell the age of starfish? Not unless you have kept track of a growing population.

189. How long do starfish live? Each species has its own average age limit. Species of *Pisaster,* for instance, may live for twenty years, while those of *Astropecten* for only five.

190. How do starfish become so widely distributed considering how slowly or little they travel? Distribution over wide territories is accomplished mainly by the free-swimming larvae which may be carried hundreds of miles by sea currents from the region where they were liberated. (See question 187.)

191. Are starfish found as fossils? Yes. Starfish of essentially the same kind as those now living occur as fossils in rocks of the Ordovician period laid down four hundred million years ago.

192. What starfish live where? *Starfish of the Atlantic Coast.* A large number of species live on the sea floor of the continental shelf at depths ranging downward from about ten fathoms and are therefore not likely to be encountered. Of those that are to be found along the shore between the tides or in shallow water a little below tide level, the majority occur from New Jersey northward. The common starfish (*Asterias forbesi*), ranges from Maine to the Gulf of Mexico and causes great destruction to oyster beds; it also feeds voraciously

Sand starfish (3–5″)

on the mussel beds along the shore, where it is more readily seen. A ten-inch span is common, while the color may be orange, bronze, purple, green, or brown. A very similar species, (*Asterias vulgaris*), the northern starfish, ranges from Cape Hatteras to Labrador. It also grows to a large size and is a menace to oyster beds. North of Cape Cod it may be found between the tides on barnacle-covered rocks beneath drapes of seaweed. Grayish purple colors predominate, but rose, red, orange and blue individuals are common. The sieve plate is light yellow, whereas that of the common starfish is a bright orange-red, which is the simplest means of distinguishing between the two species.

At very low tide in rock pools or crevices, and among the roots of kelp, from Greenland to Cape Hatteras may be found the small, smooth blood sea star, (*Henricia*) which is usually blood-red but may also be yellow; it broods its young instead of spawning freely like most other starfish.

Two sun stars occur at depths from shallow water below low tide down to considerable depths. The common sun star (*Crossaster*) known as the spiny sun star, is the more magnificent, with ten to twelve rays and often a brilliant scarlet at the center surrounded by bands of crimson, pink or white and tipped with crimson—a sunburst of color. It grows to a diameter of fourteen inches and ranges from the Arctic to New Jersey on the Atlantic Coast, south to Vancouver on the Pacific and from Scandinavia to the English Channel. The other sun star, the purple sun star (*Solaster*) ranges northward from Cape Cod. It has seven to thirteen rays and is usually of a striking but uniform red-violet color.

The only other starfish likely to be found on the northeastern coast is the mud star (*Ctenodiscus*) which occurs in shallow water on firm mud and on muddy sand bottoms as well as down to very considerable depths. The color is yellowish, the shape unmistakable for the five arms combine with the body to form a pentagon. It has a great range and occurs from the Arctic to South America.

The sand star (*Astropecten*) is common south of New Jersey and is a flat, strikingly symmetrical, sharp-rayed sea star. But the most outstanding starfish of the south is the giant starfish (*Oreaster*), often measuring twelve to sixteen inches across, found on shallow sandy bottoms from South Carolina to Florida and the Bahamas. It has a short, powerful body. A little offshore, still readily visible from a boat, these starfish congregate in large numbers on the white sand among the keys. The rays are thick and blend with the heavy body and these richly colored sea stars, deep red, or a rich maroon, orange, deep purple, are a remarkable sight seen through quiet transparent waters.

Starfish of the California Coast. The large red, yellow or purple sea bat (*Patiria*) so named because of its webbed rays, is usually the most obvious inhabitant of the tide pools and reefs of the lower part of the tidal zone. It ranges from southern Alaska to northern Mexico but is commonest in the middle region of the range. In southern California you may find the peculiar Linckia starfish on the rocky shore, noted for its voluntary fragmentation and regeneration. Linckia is slender-rayed, usually with an odd number, and relatively smooth.

The blood sea star (*Henricia*) is also present, while the colorful leather star (*Dermasterias*) is often abundant in completely sheltered bays and sounds; both are typical of more northern regions such as Puget Sound. The small six-rayed starfish (*Leptasterias*) is found near the low-tide level and in the tide pools f·rther up the shore. Possibly the largest starfish known and certainly among the most striking is the many-rayed sun star (*Pycnopodia*); specimens with a spread of more than two feet are not uncommon. It has a soft, delicate skin, usually brightly colored in pinks and purples. It starts out with six rays, but as it grows increases the number to twenty-four. It is commoner north of Monterey than to the south. A less common form is the twelve-rayed sun star (*Solaster*) with eight to fifteen arms, which lives in deeper water. The sand star (*Astropecten*) occurs on sand flats from Newport Bay to Panama, usually below the water line, half buried in the sand but occasionally exposed, sometimes with a spread of more than ten inches. This is a beautifully symmetrical starfish, the color of sand, with a beaded margin and a fringe of bristling spines.

Starfish of the Northwest Pacific, particularly of Puget Sound and Vancouver Straits. The purple starfish (*Pisaster ochraeus*) lives among the exposed rocks near low tide from Monterey Bay to Alaska. It grows up to fourteen inches and has three color phases; brown, purple and yellow. A pinch bar is generally needed to lever one away from the rock. In quieter waters of Puget Sound and British Columbia, above the low-tide level, a related form (*Pisaster confertus*) haunts the mussel beds and is mostly colored a vivid violet. Elsewhere in Puget Sound a pink-skinned, short-spined edition of the common starfish, (*Pisaster brevispinus*) attains the gigantic size of more than two feet, particularly in the Hood Canal where it lives on soft bottoms below rocky or gravelly foreshores, just below the line of low spring tides. This species also ranges south to Monterey Bay. The blood star (*Henricia*) and the six-rayed star (*Leptasterias*), both of which brood their young, also occur on the protected outer coast. The remaining form is the leather star, (*Dermasterias*), which decorated the tide pools. Its sleek skin is usually a delicate purple with red markings, with the tips of the rays turned up. It ranges from Sitka, Alaska, to Monterey, but is commonest and larger in the Puget Sound area, attaining to a diameter of ten inches. Specimens in the tide pools of the outer protected coast are generally much smaller.

193. What is the difference between a starfish and a brittle star?
In starfish the rays blend more or less indistinguishably with the central region of the body. In brittle stars the rays, always five in number, are sharply marked off from the central disc, which may be either

Spiny brittle star (4–6")

round or pentagonal. They are called serpent stars because of their long slender rays with snake-like movements, and brittle stars from the fact that the rays break very easily if the animals are handled or disturbed.

194. Where are brittle stars found? Brittle stars shun light and cannot stand exposure out of water. Accordingly they are to be found for the most part under rocks or reefs, among the holdfasts of seaweeds or in deeper water. On shallow seas they may actually carpet the sandy or gravelly floor.

195. How many rays does a brittle star have? The very great majority of species have five rays, but a small southern species, which occurs off the Florida Coast and through the West Indies, has six. This species (*Ophiactis*) also is unusual inasmuch as it often reproduces by dividing into two, followed by regeneration.

196. Why are brittle stars so brittle? Because the calcareous plates which comprise much of the flexible rays are held together by relatively little soft tissue and so break easily.

197. Can broken rays be replaced? Yes. The power of regeneration of missing parts is almost as well developed as in starfish. Rays are frequently broken either by accident or from being nipped by a crab, and are quickly replaced by growth from the stump. (See questions 171–174.)

198. How do brittle stars move? While brittle stars do have tube feet, like starfish, they are too small to serve for locomotion except in a few cases and are used mainly for respiration and sensory purposes. Most brittle stars move by extending one ray or arm out ahead, as in starfish, and push with the others, relying upon their muscular flexibility. Some species progress by using two rays for grasping objects and using the other three for pushing forward. In others one arm may be trailing and the other four curved forward and sideways so that the animal travels along the ground in a sort of hopping movement. Compared with most starfish they often move very fast. (See question 168.)

199. How do brittle stars feed? They hunt for food for the most part at night, at least in the region of the shore, and feed mainly on organic debris or on small dead animals or fragments of larger ones. The minute tube feet on the underside of the rays aid in the selection of food particles and in passing them along to the mouth at the center of the disc. They have been seen to grasp a piece of crab meat or fish flesh with the tip of an arm and then roll up in a coil so that the food is brought close to the mouth where it is eaten. While sensitive to light, they rely mainly on the sense of smell or taste which seems to be located near the ends of the rays, and the food can be detected, together with the direction in which it lies, at some distance from the animal—even several feet away.

200. What is a basket star? A basket star is a brittle star whose five rays all divide and each branch again divides and so on until very numerous terminal branches are formed. When they partly curl up, a basket appearance results. They occur on both the Atlantic and Pacific Coasts, but rarely close to the shoreline. Basket stars are often brought up on fishing lines.

201. Where are brittle stars found? Many kinds live in tremendous numbers on the open sea floor at considerable depths, but others can be found in low-level tide pools, under boulders, and particularly among the holdfasts of kelp at low tide. On the Atlantic Coast the daisy brittle star (*Ophiopholis*) with its circular disc and spiny rays of many colors is the most likely to be found north of Long Island

Sound. The disc is often red with rays banded red and white; or a blue-and-brown disc with green-and-brown banded rays, and many other combinations. A little grayish-white form with slender rays, the long-armed snake star (*Amphipholis*) may also be found, although more often under stones, whereas the daisy star is found hunting among the kelp holdfasts, and the green bread-crumb sponge lining the deeper crevices of cool rock pools.

Southward from Long Island Sound the green brittle star (*Ophioderma*), with arms almost smooth, is especially common in tide pools and among protected sand flats, its color blending with its surroundings. The spiny brittle star (*Ophiothrix*), its disc covered with bristles, of almost any color except blue, also is frequently seen. Farther south again, among the Florida Keys, you may find two of the most spectacular members of the group: the so-called sea spider (*Ophiocoma*) of the West Indies, which is large, black, bristling and very active, and lives under almost every reef and coral boulder from low tide down; and the West Indian basket star, whose scientific name, *Gorgonocephalus,* meaning the head of Gorgon, well describes it, clings with its many-branching rays and tentacles to weeds and sea fans as it fishes for food.

On the Pacific Coast the daisy brittle star, so common on the New England Coast, ranges from Monterey to Alaska, and together with the more southern brown brittle star they are among the most noticeable of under-rock inhabitants since they are both brilliant and active.

202. What are sea urchins? Sea urchins, in spite of superficial differences, are related to starfish and brittle stars. Examination of a

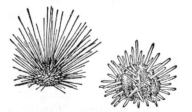

Poison-spined sea urchin (10″) (Diadema) and club-spined urchin (Cidaris) (2″)

test or shell of an urchin that has been denuded of spines will show the five-rayed design and the holes through which the tube feet protrude.

203. Are urchins edible? In spite of their appearance, urchins are highly regarded as food in certain regions. Their reproductive glands or roe are a regular article of diet along the coast of Italy, Southern France, Eastern Mediterranean. Only certain large species are gathered for this purpose. The white-spined sea urchin, or "sea egg" as it is locally known, is similarly eaten by the Negroes through the West Indies. On the Pacific Coast the large, seven-inch red urchin is considered a delicacy by the Italians. According to two biologists, E. K. Ricketts and Jack Calvin, they are very good when eaten à l'Italienne (raw) with French bread—extremely rich, and possibly more subtle than caviar.

204. Can urchins see? They are sensitive to light to much the same degree as a starfish and by the same means. A small sensitive pigment spot is located on each of five plates which surround the upper disc, where the ends of the double row of tube feet narrow to a point, corresponding to the tips of the starfish rays.

205. What is there inside an urchin? A large amount of fluid which is, in the main, like seawater; a loosely coiled gut passing from the mouth to a small anal opening in the center plate on the upper disc;

Aristotle's lantern or sea urchin jaws (1″)

five masses of reproductive glands, male or female according to the sex, suspended from the top; and the large apparatus known as Aristotle's Lantern, consisting of five teeth with their supporting and activating structures.

206. How long do urchins live? Species of sea urchins are known to live for more than eight years. The age of an urchin can be determined in somewhat the same way as one can tell the age of a fish by counting annual growth rings on the scales, or the age of a tree by its rings, only in this case with greater difficulty. If the plates of a sea urchin are carefully ground horizontally, rings of pigment are seen to alternate with limy material, the pigment being laid down at the end of each summer when shell growth has temporarily ceased and the tissues generally are loaded with pigment.

207. How does the shell of an urchin grow? Growth of the rigid shell is accomplished mainly by continually adding new calcareous substances to the outer surface of the plates, while at the same time dissolving material on the inner surface.

208. What do sea urchins feed on? Most sea urchins feed on encrusting vegetation on rocks, scraping it off with their five pointed teeth which alternately separate and come together in the mouth region. They are primarily vegetarian browsers. In many the spines serve as a trap for gathering bits of seaweed which are passed around to the mouth, bitten into small pieces by the five teeth, and swallowed. They will also eat flesh, although the opportunity rarely arises. (See question 113.)

209. How do urchins protect themselves? Spines protect them to some extent from larger animals, but more protection is generally needed against smaller forms which might become attached or even

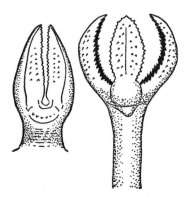

Skin pincers of starfish (1/10")
and sea urchin (1/5")

parasitic upon their outer surface, like fleas in fur, and against these both the spines and the smaller pincer-like pedicellarias are brought to bear. When an urchin's shell is prodded with a sharp instrument, the spines converge towards the point touched so as to offer a strictly mechanical defence. If, however, a blunt instrument is used, the spines turn away so that the pedicellarias at their base can have free play. The heads of these have three jaws, each provided with a poison gland, and are supported on flexible stalks. Many sea urchins cover their upper surface with pieces of weed and shells as a means of camouflage. (See question 159.)

210. Of what use are the spines? Spines have an obvious protective value, particularly where they are long and sharp, or poison-tipped, or so heavy and club-like that they are practically unbreakable. At the same time they are essential for locomotion, since they lift the globular body off the ground and are employed somewhat as stilts.

211. Can the spines of a sea urchin hurt? The spines of most sea urchins are sharp enough to penetrate the skin if carelessly handled or stepped on, and readily break off and work their way in although without serious consequences. In the case of the long-spined, black, tropical urchin which lives among the Florida reefs and throughout the West Indies, looking almost like a porcupine, the spines are slender and hollow and deliver a sting as severe as a hornet's.

212. What enemies do sea urchins have? Along the shore in shallow pools or just below low tide, urchins are taken by gulls. They are also attacked by crabs of various sorts, which use their claws somewhat in the manner of can openers to remove the tops of the urchins. Starfish are known to attack them as well. Urchins that live on the more open sea floor are preyed upon not only by crabs but by rays, certain sharks, and by seals. Along the Northwest Pacific Coast they form much of the food of the sea otter.

213. How do urchins hold their position on steep rock surfaces? By means of the suckers at the ends of their numerous tube feet.

214. How do sea urchins reproduce? All urchins shed either eggs or sperm into the surrounding water from their five gonads which open through five minute holes in the upper disc. Fertilization takes place in the water and the larvae develop and drift in the currents for many weeks before they metamorphose and settle on the sea floor. As a rule, when one individual begins to spawn in a ripe community, all the others do likewise. (See question 187.)

215. What is the shell or test of an urchin? The shell serves both as a protection and as a skeleton for the animal. It has living tissue both inside and outside and is made up of numerous calcareous plates fitted closely and accurately together to form a rigid case. There is a large hole at the center of the undersurface, where the mouth is located, and there are five double rows of minute openings radially from the mouth region to a small disc at the center of the upper surface, that is from pole to pole. These correspond to the five rays of the starfish and permit the protrusion of five double rows of tube feet. The disc on top consists of a small central plate surrounded by five others and corresponds to the disc of the brittle star or the central region of the starfish. A small hole in each of the five marginal plates permits reproductive cells to escape in due season. One of the five plates is heavier and is finely perforated all over; it is the sieve plate, which permits intake of water into the hydraulic system. The plates comprising the shell as a whole bear small knobs or bosses which serve for ball-and-socket attachment of spines. (See question 168.)

216. Where do sea urchins live? Urchins live for the most part attached to the sides or undersurfaces of rocks, ledges, boulders or coral reefs, where they feed upon the alga growths. Since the more or less globular shape of most urchins exposes them to the force of waves, they tend to occupy crevices, undersurfaces, and wherever some protection from the sweep of waves or current enables them to hold on securely.

The commonest urchins of the Atlantic and Pacific Coasts are the green urchin and the red urchin (both species of *Strongylocentrotus*). On the Pacific Coast, the giant red urchin occupies the deeper pools and the rocks from low tide down; the purple urchin, also with a diameter of six to seven inches, is more adventurous and many are found with more than half their bristling bulk sunk into rock excavations of their own making; the green urchin is more of a wanderer and rarely settles firmly in one spot. All three range at least from British Columbia to Lower California. The same green urchin also ranges south on the Atlantic Coast as far as Cape Cod and in truth is a circumpolar species. South of Cape Cod to the Virginia Capes a smaller purple form (*Arbacia*) is found in its place.

In the south along the Florida Coast, particularly among the reefs, and throughout the West Indies, interesting tropical specimens are common. The most spectacular is the very long-spined, large, black sea porcupine, (*Diadema*), with poisonous spine tips, which hides under rocks and reefs when the sun is bright and wanders out on the sandy floor on dull days and at night. Under rocks and firmly attached as though disinclined to move, you find the small club-spined urchin (*Cidaris*) which has blunt, thick but very few spines in place of the usual large number of slender pointed ones, while on the exposed sides of reefs and coral rock the boring urchins, (*Echinometra*) occupy small hollows when very young and enlarge them as they grow until the openings finally are much smaller than the urchins, making it impossible for waves to dislodge them, though restricting the urchins' food to what reaches them since they no longer can go in search.

217. What urchins live exclusively in sand or mud? Heart urchins bury themselves completely in sand or mud, while cake urchins and sand dollars live on the surface of sand or at the most only partly embedded. Heart urchins are more or less heart-shaped, in other words to some extent streamlined to facilitate burrowing. Cake urchins and sand dollars are flattened to different degrees so that they can maintain position on soft sand without danger of rolling over with every movement of the waters.

218. What is a heart urchin? A kind of sea urchin partly streamlined for burrowing in sand or mud; blunt at one end and more taper-

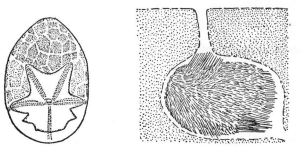

Heart urchin (2"), showing burrowing animal and shell with spines removed

ing behind. Spines are long, varying in length, point obliquely back-wards and are employed in the process of burrowing. Heart urchins feed on small particles which are gathered and passed to the mouth by their tube feet. Clean water for their hydraulic system is drawn down from the surface of the sand and mud through a tube.

219. Where do heart urchins live? The cosmopolitan golden heart urchin (*Echinocardium*), which is commonly found buried in clean sand just at low-tide level down to deeper water along the coast of Europe, is typical of its kind. Its presence along the sandy shores of North America is suspected but not certain. Other than this, the southern heart urchin (*Moira*) is common in shallow water from North Carolina southward to the West Indies, buried in soft mud, while on the Pacific Coast a red to rose-lavender species (*Lovenia*) occurs on the southern Californian sand flats where it lies half buried near extreme low tide.

220. What are sand dollars and cake urchins? Sand dollars are ex-tremely flattened and cake urchins (also called sea biscuits) are partly flattened forms of sea urchins adapted to living on the open sea floor, particularly on the surface of sand or muddy sand. Where sand is soft, both the sand dollars and cake urchins are usually

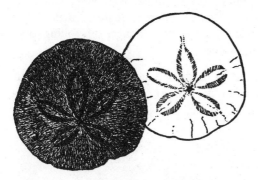

Sand dollar (2–3″) with and without covering of spines

slightly embedded so that their margins are submerged, but the upper central surface remains exposed. In both types the spines are very small and very numerous, forming an almost velvety coat. The dead shells denuded of spines are often found tossed up on shore by the tides.

221. What are the five petal-shaped areas on the upper surface of the sand dollar and cake urchin? They are regions perforated to permit the protrusion of small respiratory tube feet.

222. What are the five grooves on the undersurface of a sand dollar? They unite with grooves on the upper surface and serve to pass strings of mucus or slime towards the mouth at the center of the undersurface, and carry food particles collected from the water into the mouth.

223. How do sand dollars reproduce? In the same manner as sea urchins. (See question 214.)

224. How do sand dollars move? While the hydraulic tube-feet system is present, as in starfish and sea urchins (see question 187) locomotion is the result of waves of co-ordinated spine movements.

225. What are the enemies of sand dollars? Starfish are the principal enemies of sand dollars, as well as of sea urchins generally. When a starfish crosses a bed of sand dollars, those immediately downcurrent from it usually sense its presence and quickly burrow beneath the surface.

226. What rattles when you shake the dead shell of a sand dollar? The five teeth which form the main part of the Aristotle's Lantern or masticatory apparatus, which is like that of the sea urchin although much flattened. (See questions 205, 208.)

227. What are the slots for in keyhole sand dollars? They strengthen the shell by increasing fusion between upper and lower shells and also

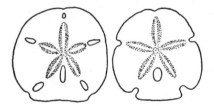

Shells of keyhole sand dollars (2–3″)

serve to allow sand and other particles to sift from the lower to upper surface, thereby aiding movement and covering the upper surface with camouflaging sand.

228. Where are sand dollars and cake urchins found? The more or less round, dark purplish-brown sand dollar (*Echinarachnis*) is generally abundant on firm sand in sheltered water northward from Long Island Sound on the Atlantic Coast and north from Puget Sound on the Pacific. Another deep-purple form, (*Dendraster*), with its star-design somewhat off-center, occurs in great beds in sheltered waters from Puget Sound south to San Diego Bay, in places so dense that up to five hundred may be found within one square yard. Along the more southerly part of the Atlantic Coast, especially from Cape Hatteras south into the West Indies, two sand-colored forms live partly submerged in the extensive sand flats exposed at low tide; both have slots or keyholes in the shell, one of them, (*Mellita*), with five slots set completely inside the margin and one (*Encope*) with six slots, most of which extend into the edge.

The thicker cake urchins (*Clypeaster* and *Echinanthrus*) with size and proportions not unlike those of hamburger buns, as distinct from the wafer-thin sand dollars, live on sandy bottoms, partly hidden by sand sifting onto their upper surface, from the Carolinas south and along the Gulf Coast.

229. What is a sea cucumber? Sea cucumbers seem, at least at first glance, to have little in common with starfish and sea urchins, yet they have the same fundamental five-spoke pattern. If viewed from one end, the design can be recognized. Sea cucumbers are in fact soft-

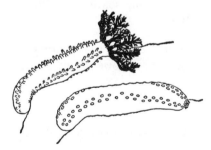

Sea cucumber, expanded and contracted
(1/2–15″)

bodied, elongated echinoderms, as though a sea urchin had dissolved its calcareous plates and had become drawn out between its upper and lower surfaces. In some species five rows of tube feet are evident; in others only three rows remain; and in some the whole surface may be covered with them.

230. Where do sea cucumbers live? The majority of species live on rocky shores, attached to rock surfaces by their tube feet, usually in crevices, under rocks or on ledges, shallow water or near low tide level. In deeper water they may lie on exposed rocky floor. Others live on the surface of sand or mud flats in very sheltered waters such as lagoons or inlets, while a more specialized kind burrows in sand or sandy mud and is never seen at the surface.

231. Which is the front end of a sea cucumber? The end through which a crown of thick tentacles protrudes. If the animal is contracted and the tentacles are withdrawn you may have to put it in a pool and wait for a long time to see them. Otherwise it is difficult to tell which end is which.

232. What are the tentacles for? For feeding.

233. Where is the mouth? At the center of the ring of tentacles.

234. How does it feed? All sea cucumbers feed by means of their tentacles, but in somewhat different ways according to circumstances. Those living attached to rocks feed on microscopic plants, particularly diatoms, collected from the surrounding water. Those that live on sand or mud flats feed by sweeping the surface with their tentacles. Those that actually burrow in sand or sandy mud also employ their tentacles, but swallow much of the sand in the process, somewhat as an earthworm swallows soil. In every case the tentacles are covered with a sticky slime to which particles adhere, and the animal puts one tentacle after another into its mouth and scrapes off whatever has become stuck to it.

235. What does it eat? Microscopic plant and animal organisms collected from the water, from the surface of sand or mud, or below the surface, each cucumber according to its kind.

236. Why does a sea cucumber withdraw all its tentacles? Either because it has been disturbed and withdraws the feeding structures for the sake of safety, or because it may be already congested with food.

237. Does a sea cucumber have any senses? All sea cucumbers have a well developed sense of touch, but are relatively insensitive to light.

238. How do sea cucumbers breathe? Burrowing forms breathe through the skin but the other kinds have a special system. They have two large, branching, tree-like sets of tubes inside the body, the trunks of which open into a large cavity just inside the anal opening of the animal, at the end opposite to the tentacles. By gulping in water through the anus and filling this system with water they are able to bring oxygen to all parts. Water is then squirted out and a new supply taken in. The process is easily seen.

239. Is their shape an advantage? Since most sea cucumbers are typically inhabitants of rocks their shape and flexibility enables them to squeeze under rocks and ledges and into crevices where they are protected from predators and from the force of waves or currents which might otherwise dislodge them.

240. How does the sea cucumber move around? By means of rows of tube feet, except in the case of the burrowing forms, which have lost most of their tube feet and progress by muscular movements of the body in the manner of burrowing worms.

241. What are its enemies? Fish, and crabs to some extent.

242. What are the sea cucumber's means of protection? A tough leathery skin in most cases. In some forms, such as species of Stichopus, warty glands on the surface exude tough, sticky threads of a repellent nature, while in many cases the animal contracts and forces out its internal viscera, which usually satisfies the attacker, leaving the empty leathery case to regenerate a new inside. Sand-burrowing forms usually break into two or more pieces when roughly handled and each regenerates the missing part if given a chance. (See questions 171, 197.)

243. Are sea cucumbers ever used as food? A large southern form (*Stichopus*) which has a more glutinous and less leathery body wall than other forms is highly prized by the Chinese. After being boiled and dried it is sold by Chinese merchants everywhere and is known as "trepang" or "beche-de-mer."

244. What colors are found? The common forms along the Pacific Coast and the Northern part of the New England Coast are reddish brown. Others may be almost black, some nearly white, while the sand-burrowing forms are usually more or less sand-colored. One form with brown scales on one side has vivid rose or purple tentacles.

245. What kinds of sea cucumbers are there? Those that look most like cucumbers (species of *Cucumaria*) live attached to rocks if they are large kinds or to weeds if small. The northern sea cucumber of the Atlantic Coast and the red sea cucumber of the Pacific Coast are both large reddish-brown forms six to ten inches long, to be found under rocks and ledges at extreme low tide. Smaller Cucumarians may be found attached to eelgrass in the muddy coves of North Carolina and on the encrusting red seaweed near the low water level in Monterey Bay. The scaly cucumber (*Psolus*), about six inches long, with a flat side attached to the rock surface and its upper side cov-

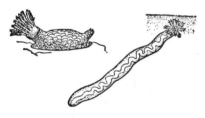

Unusual sea cucumbers: the scaly cucumber (3–4") and the burrowing synapta (6–18")

ered with overlapping yellow-brown scales, is a fantastic-looking creature found at extreme low tides along the northern New England Coast and the northwest Pacific Coast. A peculiar kind (*Thyone*), with tube feet scattered thickly all over the body, is common in shallow muddy water from Vineyard Sound southward. Then there are the burrowing forms (*Leptosynapta*), which are pale, worm-like creatures that burrow in firm sand or mud where it is left by the receding tide on both the east and west coasts. Lastly there is the warty "trepang" (*Stichopus*), a foot or more in length, which also occurs on both coasts, though in southern regions only, and is the kind used in the Orient as food.

MOLLUSKS—GASTROPODS

246. What are gastropods? The word means "stomach-footed." They
are the sea snails, whelks, limpets and conchs. The name applies to
all mollusks which crawl on a broad-based foot and have a single
shell, including those where the shell has been lost as in the case of
slugs. They have a distinct head, usually with eyes and tentacles. The
shell is typically spiral and the body of the animal is in other ways
adjusted to the nature of the shell. (See question 148.)

247. Do all snail-like mollusks have shells? Sea slugs, as well as
land slugs, have no shells once they start to crawl about, although
they start with a microscopic shell in the usual way, only to discard
it as growth continues. Sea hares and bubble snails on the other hand
only appear to have no shell, for the sides of the body grow up over
the top and hide it completely from sight. Their shells no longer having much protective value are generally light and tend to be tossed
high on the beach.

248. What senses do sea snails have? They have an acute chemical
sense akin to smell and taste. They have one or two tentacles sensitive to touch, as is the skin as a whole. In most cases but not all, a
pair of eyes equipped with lens and retina are also present, although
they are necessarily so short-sighted, probably for distances of a fraction of an inch, that they serve more for the detection of changes in
light intensity than for forming images. Snails also have a pair of
gravity-sensitive organs embedded in the brain tissue.

249. How do sea snails breathe? Water is drawn in through a
funnel-like fold in the mantle into the gill chamber where the gills
extract the oxygen except in the case of the sea slugs, where the
absence of the shell has made other devices possible.

250. What is an operculum? The operculum is the horny shield
seen on the under side of the rear end of a gastropod and in most
cases is used only to close the opening of the shell when the animal
is withdrawn. In certain conchs, however, it is used as a claw as an
aid to locomotion. (See questions 266, 277.)

Waved whelk (3–3 1/2") showing closing of shell opening by the operculum

251. How do sea snails move? By means of almost invisible waves of muscular movement which pass along the length of the foot and cause it to move forward.

252. What is a radula? The radula is a rasp-like ribbon of small horny teeth that all gastropods and certain other mollusks possess for rasping either flesh or vegetable matter and conveying it into the mouth. It is also used for drilling holes in bivalve shells. (See question 152.)

253. How do sea snails breed? The great majority of them are of separate sex. A few like the slipper limpet are first male and then, as they grow further, become female. Some others, such as the sea hare, are full hermaphrodites, that is, are male and female simultaneously and double-mate with similar individuals.

Eggs are never shed freely into the water but are either laid in large numbers embedded in some kind of gelatinous spawn usually fastened to rocks or weed or even as a raft of bubbles trapped in slime, or else are laid in firmly anchored parchment-like capsules. In the former case, the eggs are comparatively small and hatch out as free-swimming larvae called veligers, which settle on the sea floor usually after days of drifting life; while in the latter case the eggs are usually yolky and hatch out as miniatures of their parents, already equipped to crawl and feed on the sea bottom.

254. What are whelks and conchs? They are carnivorous snails, which feed mainly on bivalves and any dead flesh. The smallest species, less than one inch long, are generally called dog whelks. Larger species are known either as whelks or conchs. Yet, except in so far as their habits must conform to their size, they are essentially alike. (See question 256.)

255. Are whelks edible? Probably most of them are, as are most mollusks, but only the waved whelk is actually eaten. This is the edible whelk of northern Europe and the British Isles, where it has been a regular article of food for centuries. It is just as common on the northeastern American Atlantic Coast but is merely regarded as a nuisance by fishermen who find them eating the bait in lobster traps or on cod lines.

256. Is there any distinction between a whelk and a conch? The names are rather loosely employed and are to some extent interchangeable, but the whelks proper have a narrow projecting funnel-like extension which in life contains the siphon, whereas a conch usually has a broad flaring lip to its shell without the obvious siphon cover. One outstanding exception is the horse conch, which would be called a whelk were it not so enormous.

257. What is the long projecting part of the shell of a whelk for? It serves as a protection for the long tubular siphon through which the animal draws in water to its gill without interfering with or being contaminated by feeding activities.

258. Why are dog whelks so called? Not for any resemblance to dogs, but because they are small whelks and the prefix "dog" is an old diminutive.

259. What are dog whelks? Apart from the rock purple dog whelk (see question 261) which preys on the barnacles and mussels on the tidal rocks, the dog whelks as a group constitute the most abundant snail-like mollusks along the shore. They are individually small and inconspicuous, but abound in pools and shallow places, particularly in muddy or sandy situations.

260. What do dog whelks feed on? They are possibly the most effective scavengers of the shallow waters. A dead fish or a crushed clam thrown into the water will attract hundreds. "One morning at Morro Bay, Calif., when the tide was running out, we saw a dead fish lying in a narrow channel. For a distance of at least 100 feet below the fish,

numbers of Nassarius, having scented the food, were plowing up-
stream, while many others were already busy at the feast. . . . In the
laboratory these snails have taken hold of a piece of meat and hung
on so tightly that they allowed themselves to be lifted clear of the
water before dropping loose."—(G. E. and N. MacGinitie.)

261. Why are some dog whelks called purples? The purples are so
named not for the color of their shell but because they are reputed
to be the sea snail from which purple dye of the ancient world was de-
rived. Their anal glands secrete a purple or crimson dye which was
made use of not only by American Indian tribes but also by the Phoe-
nicians, who combined it with the dye obtained from the related sea
snail Murex to produce certain varieties of the famous Tyrian purple
used for the robes of emperors and kings. The dye is poisonous and
the whelks may use it to paralyze the muscles of their victims, or perhaps to make themselves unappetizing to others.

262. Where are rock purples found? Primarily they live on the rocks
and in pools of the intertidal region, for they are fundamentally dwarf
whelks which have come ashore to exploit the abundant food to be
found there. Larger whelks would have difficulty in resisting wave action. The most spectacular species is the rock purple (*Thais lapillus*)
common on the New England Coast and the shores of Europe. The
Florida dye shell (*Thais floridana*), ranging from Cape Hatteras to
Texas, and the short-spined purple (*Thais emarginata*) of the Pacific
Coast are less colorful as shells but basically the same animals. Most
species inhabit open rocky shores exposed to the force of breaking
waves.

263. What do rock purples feed on? Their principal food comes from
two animals that cannot move away if attacked—namely, barnacles
and mussels. The browsing snails of some rocky regions, such as the
limpets, periwinkles and top shells, are also preyed upon.

264. How do dog whelks feed? The dog whelk has a proboscis which
can be protruded for some distance. At the end of it is the usual rasp
or radula, exceptional in that the rasp teeth are few and prominent.
With this it slowly drills a narrow hole through the shell of the victim,
after which the proboscis extends still farther and rasps out the flesh,
passing it back as on a conveyor belt to the throat. In the case of

barnacles the tedious process of boring is unnecessary and the barnacle valves are simply forced apart and the proboscis inserted. It is possible that the purple dye, called purpurin, may first kill the barnacle, causing its muscles to relax, for the dye is certainly highly poisonous.

265. Why are the shells of the rock purple dog whelk so variable? Shells may be pure white, banded black-and-white or brown-and-white, or yellow or orange all over. The colors reflect to a great extent the diet of the individual throughout its period of growth. Brown-black and mauve-pink individuals have fed mainly on the rich reddish-brown flesh of mussels, white ones exclusively on barnacles. Those with striking bands of black and white have made abrupt changes from mussel to barnacle and back again. Yellow and orange forms are barnacle feeders exposed to excessive wave action, although why they should be yellowish is obscure.

266. How do rock purples manage to live so long out of water while the tide is out? By seeking the cooler, damper crevices and by pulling the shell down against the surface of the rock. An operculum is present but is rarely employed to close the shell unless the animal loses its grip.

267. What are the numerous straw-colored capsules, shaped like grains of wheat, found in crevices and under boulders? They are the egg capsules and are laid throughout the year, although more

Rock purple or dog whelk (1") alongside egg cases

particularly in late winter and early spring. The whelks have to pair before the eggs are laid, and for this purpose collect in large numbers together, usually in crevices. The animal takes about one hour to lay a single case and produces from six to thirty-six in one season, although by successive spawning may lay between 200 and 300.

268. How many eggs are there in an egg capsule of a rock purple? To begin with, there are several hundred comparatively large, yolky eggs, but the great majority of these are not fertilized and serve as food for the few that do develop. After about four months some ten or twelve small, fully shelled individuals emerge and start searching for food, although lower down the shore than the rocks where they were born.

269. What do very young rock purples feed on? When they first emerge from the egg capsules the tiny dog whelks are carried down the shore by the retreating tide, and there they find food better suited to their size. This consists of the small tube-worm *Spirorbis* which lies in coiled limy tubes. On this the young whelks feed until about one third of an inch long after which they move up the shore to the barnacle zone.

270. Where is the waved whelk found? The waved whelk (*Buccinum*), three inches long, lives along the Atlantic Coast as far south as New Jersey but also extends through northern coastal waters generally and is common around the British Isles. It can be found on mussel beds at dead low water where tidal currents are strong, or by tying a dead fish in a bag of loose netting and anchoring it among the rocks near the low water line.

271. What are the egg masses of the waved whelk? The whole egg mass of a single whelk forms a loose ball about the size of the palm of your hand, each mass consisting of scores of capsules about the size and shape of a split pea. Each capsule contains hundreds of eggs

Egg case mass of waved whelk (2–3″)

and in due course they hatch and escape from the capsules as minute but fully equipped editions of their parents, to crawl about in search of equally minute bivalves and worms to conquer. Sailors once used the egg masses to wash their hands, so that they are sometimes called wash balls. Empty balls are often found along the beach at high tide.

272. Why are left-handed whelks so called? When a snail or whelk is viewed from the tip or center of its shell, the shell is seen to follow a spiral course. In the vast majority of cases the spiral is right-handed, as in most of the whelks, for instance, found along the Atlantic Coast. The left-handed whelk is one of the exceptions in which the spiral turns to the left instead of to the right.

273. What are the egg masses of Neptune whelks like? The giant knobbed whelk, the channeled whelk, and the left-handed whelk lay peculiar egg strings which are familiar to most persons who stroll the beaches of the Middle Atlantic States. The eggs are laid in double-

Channeled whelk (7″ shell) with string of egg cases

edged, parchment-like and disc-shaped capsules an inch or so across, attached together on one edge by a cord of the same substance. The string may have as many as one hundred capsules and may be a yard or more in length. The empty strings washed up by the tides are an arresting sight.

274. What do conchs eat? Most of the larger conchs, such as the queen and the fighting conchs, are carrion feeders and are drawn toward any dead fish or other creature. The giant horse conch, however, feeds on cask shells, while the smaller crown conch feeds mainly on bivalves.

275. What kind of mollusk is the crown conch? It is one of the smaller conchs (*Melongea*) from two to five inches long, with a claw-like operculum. According to Percy Morris, "This is an active, predatory gastropod, well able to take care of itself in a sea of enemies. It is said that the giant horse conch (*Fasciolaria gigantea*) is the only mollusk capable of overpowering it. It feeds largely on bivalves, coon oysters being perhaps its favorite food, but it does not hesitate to attack and devour *Busycon* and other snails. Observers have reported what appears to be signs of animal intelligence on the part of these carnivores. They have been noted circling around in back of a resting scallop (*Pecten*) in order to sneak up from the rear, approaching within pouncing distance before the agile bivalve is aware of danger. The crowned conch prefers muddy and brackish water, and is common throughout Florida."

276. How does the large queen conch move? This is the largest of the carrion-eating conchs and it moves its massive body and heavy shell over the sea bottom in a series of grotesque leaps, tumbling over in the process. It is primarily a scavenger, like the fighting conch.

277. Which is the fighting conch? The fighting conchs include both the fighting conch itself, which is common in the West Indies and rare in Florida, and the Florida conch which may be found in numbers feeding upon dead fish in shallow water. They are pugnacious active mollusks which make good use of their claw-like operculum in moving about when stranded by the tide. The claw is hooked into the sand and the shell raised high in the air, toppling over successively so that an awkward but effective progress is made toward the water. They fight mainly to retain their food from other hungry conchs, making violent movements of their foot and shell.

278. Why is the horse conch so named? Just as "dog" is used to denote comparatively small kinds of an animal, "horse" implies unusually large, as in the case of the horse mussel. The horse conch is the giant among the large conchs. The shell is heavy and massive and grows to two feet in length, and is found from North Carolina to Brazil. It overpowers and smothers its prey by sheer weight and strength and is said to be the only mollusk capable of successful battle with a crowned conch. Horse conchs feed exclusively on the large cask shells.

279. Is any use made of conchs? Mainly the queen conch. "This is the shell that for generations has been used as a doorstop, or for decorating the borders of flower-beds, by families of seafaring men. It is one of the largest and heaviest gastropods, individuals sometimes weighing more than five pounds. . . . It is a commercial shell, and large numbers are exported from the Bahamas for cutting into cameos, the scrap material being ground to powder for manufacturing porcelain. Its flesh is eaten in the West Indies, the aborigines formerly making scrapers, chisels and various other tools from its shell. Semiprecious pearls are occasionally found within the mantle fold."— (Percy Morris.) The large helmet conchs, often called cameos, have been used extensively for making cameos. They are found along the Carolinas and Florida coasts.

280. How are cameos made? The conch shell is cut so that the outer white layer stands in relief against the darker inner layer. The sardonyx helmet, sometimes called the black helmet, is a preferred one for cameo cutting, as there is a very dark coat beneath the outer layer, so that the figure stands out well against the onyx background. Most of the helmet shells and other conchs yield a cameo with a reddish orange or pink background.

281. How are conch horns made? The conical tip of the shell is cut off, leaving an opening about three quarters of an inch across. The shell should be eight or ten inches long. It is then blown like a bugle. They have been used from time immemorial by fishermen to signal to each other and to tell their waiting wives of their returning home.

282. When you put a conch or other large shell to your ear, what do you hear? You do *not* hear the sound of distant surf, as the legend says, but a roaring sound produced by vibrations in the columns of air within the shell, caused by the changing pressure of your blood.

283. What is notable about the Florida cask shell? The animal itself (*Tonna galla*) is very large and seems much too big for its shell. The shell is thin and light, but is surprisingly strong for its weight. The mollusk feeds for the most part on sea urchins but is itself a frequent victim of the crown conch.

284. Where do bubble-shell snails live? On the Pacific Coast, according to Ricketts and Calvin: "In the south a large tectibranch, the bubble-shell snail, *Bullaria gouldiana* with a shell some two inches long, is the commonest of all gastropods in the Newport Bay and Mission Bay region—so common that whoever has been there must surely have seen it. The paper-thin shell is mottled brown, and the body, which is too large for the shell, is yellow. When the animal is completely extended and crawling about in the mud, the mantle covers most of the shell. . . . The long strings of eggs, lying about on the mud or tangled in eelgrass, are familiar sights in summer." On the Atlantic Coast a similar but smaller bubble-shell (*Bulla occidentalis*), about one inch long, inhabits sandy mud flats, the muddy banks of river mouths and brackish water generally, concealing themselves under seaweed or in the mud when the tide is out. It ranges from Florida to Texas.

285. What do bubble-shell snails feed on? The eastern bubble-shell feeds upon small mollusks which it swallows whole and breaks up by means of interior calcareous teeth within the stomach. The West Coast form is also reported to swallow bivalves and smaller snails whole, but the MacGinities record that it also ingests the surface film of mud for its contained nourishment.

286. What marine snails form an important part of wild ducks' diet? Two in particular both about one half inch long, namely, the salt-marsh snail (*Melampus bidentatus*) common on the grass of salt marshes overflowed at half tide, from Nova Scotia to Texas, and the coffee-bean shell (*Melampus coffeus*) which may be seen on mud flats in southern Florida, or high up on grasses and low shrubs at the full tide. Both kinds form a very important food supply for wild ducks.

287. What is the oyster drill? A very small whelk (*Urosalpinx*) which after the starfish is the worst enemy of the oyster beds. Scores of these little drills mount a clump of oysters and drill neat little pin holes through the shells, although they are even more devastating in the attacks on very young oysters. It is one of the commonest sea snails on the Atlantic Coast and unfortunately has been introduced to the Pacific Coast and to the coast of Europe where it has become as much of a pest on the oyster beds as it is here.

288. What are cowries and where are they found? A large family of mainly tropical forms with a highly distinctive oval shell with toothed lips running the whole length of the shell, with richly colored representatives distributed all around the world. A few species live as far north as Florida—namely, the large and handsome measled cowry, from three to four inches long and purplish brown with white rounded spots; the small gray and the little yellow cowry about an inch long (all species of *Cypraea*), and the diminutive coffee beans (species of *Trivia*). The small nut-brown cowry (*Zonaria*) also may be found along southern California. Cowries have long been used as a form of money by primitive tribes, particularly in Africa.

289. Why are cowry shells so highly polished? Because the mantle tissue which underlies the shells of mollusks can here be expanded so that it extends over the outside surface of the shell as well, until the mantle on the two sides meets on the top, almost hiding the shell altogether. The presence of the mantle maintains the polish.

290. What do cowries feed on? The large measled cowry and the four-spotted coffee bean are scavengers of dead animal matter. The small pink cowry common along the southern coast of England not only feeds exclusively on compound sea squirts, in particular on the golden star sea squirt (*Botryllus*), but lays its eggs in capsules imbedded in the sea-squirt jelly. The nut-brown cowry of the Pacific Coast has been seen by the MacGinities to feed on compound sea squirts, but also on small sea anemones, egg capsules of other snails, sponges, the algæ film on rocks, and even on dead abalones.

291. What is the magpie shell? One of the most striking shells found along the Florida coast, although never with the living mollusk inside, is the magpie shell (*Livona*). Three to five inches across and black with heavy splashes of white, it is to be found alive throughout the West Indies on weedy bottoms in shallow water and at the surface of reefs. The shell is very pearly and takes a high polish when the outer layer has been rubbed or peeled off. In Florida it is frequently found in old Indian shell heaps, suggesting that at one time it was abundant along that coast.

292. What are the shells that are shaped like worms? Many, of course, are actually tubes of lime made by worms but some which look like calcareous worm tubes belong to gastropod mollusks. Three kinds are to be found along the Atlantic Coast in shallow water, particularly in the southern regions: the usually solitary Serpulorbis found attached to stones and shells, and the more tangled masses of the

Worm-tube snail (3"), showing pulled-out spiral of solitary shell and slime nets sent out by feeding individuals in cluster

worm-tube shell (*Vermicularia*) whose shells still show their spiral molluscan character at their pointed end, and the reef-tube shell (*Vermetus*) which sometimes forms masses so large as to constitute reefs by themselves. On the Pacific Coast a similar tube mollusk (*Aletes*) ranges from Monterey to Peru, beneath rocks and attached to wharf piles.

293. How do the shells of the worm-shell mollusks get their shape? The young stages have typical snail-like, coiled shells. It is only after the young have attached and started to grow that the shell loses all resemblance to a snail's. New shell material is added to the edge of the shell to produce a long, worm-like tube that is irregularly coiled almost straight.

294. How do worm-shell snails feed? "*Aletes* makes use of mucus to trap its food. From a gland near its mouth it secretes mucus that extends upward in the water as a triangular sheet. This sheet of mucus is allowed to float and wave in the water for a while, then the animal pulls it down and eats it with what food material has adhered to it. . . . When they occur in clusters, the fan-shaped sheets of mucus they put out become entangled, and the table with its bill of fare becomes a community affair. When one individual in such a group begins to eat the mucus sheet all the other members start swallowing."—(G. E. and N. MacGinitie.)

295. Can a worm-shell snail move about? Only by extending and retracting its head. The body is anchored within the shell as in the case of all mollusks, while the shell itself is cemented to rock or piling or other worm shells when minute, and is never moved again.

296. How can you distinguish between a molluscan worm tube and a real worm tube? The small pointed end of a molluscan tube usually shows something of its spiral nature, which no true worm tube does. If the shells are occupied, it is a fairly simple matter to tell which kind is which, for if inhabited by worms the heads snap back in a flash with no more than a shadow to alarm them, while if inhabited by a mollusk the withdrawal is slow, more like that of a retracting sea anemone. (See question 257.)

297. What are periwinkles? Periwinkles are species of small, usually dingy, shore snails of the genus Littorina, which is Latin for "shore-dwellers," and are accordingly also known as littorines.

298. Are periwinkles edible? The common periwinkle (*Littorina littorea*) is used extensively as food. "This species might be called the English sparrow of the mollusks. Probably the most abundant and best-known snail on the North Atlantic Coast, this periwinkle is an immigrant from European waters, having been introduced (accidentally, probably in the egg stage) in the vicinity of Nova Scotia. . . . In Europe these dingy little snails are roasted and sold from push carts in the city streets. In America there seems to be an aversion to eating them, though just why we should consider clams, oysters, and scallops as delicacies and at the same time be horrified at the thought of eating a snail, it is a little hard to figure out. Anyone who through necessity or curiosity, has tasted them, pronounces them excellent, and the writer can testify that they are equal in flavor and texture to any other mollusk."—(Percy Morris.)

299. What do periwinkles feed on? All periwinkles are vegetarian and browse on rocks and seaweeds, scraping off the small growths with their rasping, ribbon-like radula.

300. To what extent do periwinkles move around? Tagging experiments on periwinkles indicate that they are great stay-at-homes, rarely migrating from the immediate neighborhood of the pool in which they happen to have been born.

301. How long can periwinkles live out of the sea? It depends on the species. Those typical of the upper regions of the intertidal shore have been kept dry experimentally for forty-two days without being damaged, and can stand immersion in fresh water, which ordinarily kills marine animals, for eleven days.

302. Is it possible to drown a periwinkle? Most periwinkles are truly marine, are active under water and merely survive while the tide leaves them and returns. Others such as the rough periwinkle of the Atlantic Coast have a reduced gill and employ the gill chamber more in the fashion of a lung for obtaining oxygen from air rather than water. To keep the breathing chamber moist, these require occasional submersion such as would occur every two weeks at the high spring tides, but permanent submersion can drown them.

303. How do periwinkles survive out of water? By withdrawing into the shell and closing the tightly fitting horny operculum, which serves to retain moisture in the gills and to keep out fresh water or drying wind. (See question 266.)

304. How do periwinkles breed? The different species breed in different ways according to where they belong between high and low-tide levels on the shore. Each has an interest of its own.

305. How does the common periwinkle breed? The common periwinkle, like the limpet and chiton, sheds its eggs and sperm freely and independently into the sea when the tide is high and the moon is in the right phase. Eggs are produced in very large numbers, and while most of them are swept away by currents, never to return, many from their own and other localities settle on the shore after many days of drifting in the sea.

306. How does the rough periwinkle breed? In these winkles there is actual copulation, the eggs are fertilized while still within the mother and wrapped around with an egg case there and then. And to make doubly sure, the case is retained within the mantle chamber of the mother until the young are able to crawl out and start feeding beside her. The call of distant places is ignored and the rough periwinkle family is brought up domestically in its own back yard.

307. How does the smooth periwinkle breed? The smooth periwinkle lays fewer, though larger, eggs and instead of throwing the precious cargo to the high seas, encloses them in small masses of jelly fastened to the weeds. Of course, when the tide goes down the egg masses may be killed by heat or drought, but those that do survive and grow are at least in the right place, and the hazards of chance dispersal are eliminated. Also, eggs that are laid in a mass must be fertilized where they are spawned, and the smooth periwinkles pair off in their own smooth periwinkle way.

308. Where are periwinkles found? Along most shores but especially those with weed-covered rocks. Each species has its own particularly vertical range. On the shores both of England and New England the common periwinkle, (*Littorina littorea*) which is the largest and has a black or brownish shell, lives for the most part but not exclusively on the lower half of the shore, the smooth periwinkle (*Littorina littoralis,* also known as *Littorina obtusata*) is confined to the seaweed zone (seaweeds *Fucus vesiculosus* and *Ascophyllum*) and usually has a bright yellow shell, while the rough periwinkle (*Littorina rudis,* also known as *Littorina saxatilis*) inhabits the higher region from half-tide to the splash zone. Farther to the south on the Atlantic Coast other periwinkles replace the above species. The gulf periwinkle (*Littorina irrorata*) lives among seaweeds along the shore, the southern periwinkle (*Littorina angulifera*) clusters on wharf pilings and the roots and leaves of mangroves, while the zebra periwinkle (*Littorina ziczac*) may be seen by the shore on rocks at low tide in southern Florida. Pacific Coast species are equally selective, with *Littorina planaxis* common on the higher rocks and *Littorina scutulata* from the middle region down to low water. Altogether the littorines, which is a more diagnostic name than periwinkle, constitute one of nature's experiments in the progressive colonization of the intertidal region. In fact, some tropical species have taken to living permanently out of water among land vegetation.

309. Where do the violet snails (or purple shells) come from? The delicate lavender or purple shells shaped very much like those of land snails, which are frequently found along the Florida and Carolina shores, belong to the floating, eyeless violet snail (*Janthina*) which is found in practically all warm seas. The snail forms a remarkable raft or float of air bubbles trapped in viscous mucus and attaches its egg capsules to the underside so that it has no need to come ashore or go to the sea floor in order to lay. It usually occurs in large schools far from land but is sometimes brought inshore by wind and currents.

310. What kinds of limpets are there? There are the true limpets with intact, rather flat conical shells without traces of spiral or a terminal opening, shaped rather like a Chinese hat, and there are the so-called keyhole limpets which have a slot-shaped hole at the apex of the shell and are more closely allied to the great abalone than to the true limpets.

311. Where are limpets found? Of the limpets, the large owl limpet (*Lottia*), more than three inches long, lives attached to surf-swept rocks in the middle tidal zone from northern California to Mexico. The smaller limpets (various species of *Acmaea*) are found in a

Limpets and keyhole limpet (1/2–1″)

Chaffy limpets attached
to blades of eel grass (3/4″)

variety of habitats; the common Pacific limpet, an inch or so long, often occurs in great colonies on rocks even where the surf is strong, and clear up to the splash line, frequently accompanied by the brown and white shield limpet and the ribbed limpet: the little chaffy limpet lives attached to the narrow blades of eelgrass even in the face of considerable surf. On the east coast the tortoise-shell limpet is common in tide pools and on the sides and undersurfaces of rocks from Labrador to Connecticut, with a variety of the same species living attached to eelgrass; the southern limpet is generally similar in looks and habits, and occurs in southern Florida. The common limpet of European coasts (*Patella*) perhaps should be mentioned since it has been more closely investigated than any other limpet.

312. Do limpets move around? Generally by night, particularly when covered by the tide, when the foot extends and lifts the shell from the rock, to go browsing in search of food. At the most, however, they travel about three feet from home and usually very much less. They appear to have some sort of "homing" instinct or mechanism, which however does not seem to be based on their limited power of sight, smell and touch. But by dawn or a little later, nearly every limpet will return to the one spot where its shell fits exactly the rock below.

313. How do limpets hold so tightly to a rock? By means of the flat, rounded, muscular foot, although the mechanism is still somewhat of a mystery. If it acts as a sucker, it must be as a number of small independent sucker areas rather than as a single large one, for otherwise a limpet could not grip small pieces of rock as it does. It

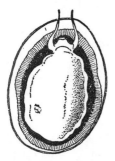

Underside of limpet (1/2–1″) showing muscular foot with surrounding mantle

takes a knife blade or a sudden knock to dislodge one; not even the stormiest seas can displace limpets from their rock surfaces. The shape of the shell and the power of the foot are indomitable. Once the foot has taken hold, it requires a seventy-pound pull to remove a limpet having a basal area of one square inch. (See question 345.)

314. How do limpets survive on rocks while the tide is out? By pulling the margins of the shell down against the rock so tightly that water is retained in the narrow groove between the central foot and the margin of the shell.

315. How do limpets fit themselves to the irregular surface of a rock? "The rock surface is seldom perfectly flat and yet the margin of the shell must make perfect contact with it if water is to be retained when the tide falls. To accomplish this the limpet has either to grind the surface of the rock to fit the margin of the shell or else make the margin fit the irregularities. It may do either of these things. If an exposed limpet be dislodged from a soft rock it will be found to have occupied a ring-shaped depression on the surface of this."—(C. M. Yonge.) If the rock is softer than the shell the rock is thus made to fit the shell, but if the rock is harder than the shell the same grinding movement wears the shell down to fit the rock.

316. What do limpets feed on? Some limpets feed on the larger seaweeds, but most browse on the young microscopic stages of seaweeds that are forever starting to grow on rock surfaces. It is estimated that for every square inch of limpet about seventy-five square inches of encrusting weed are necessary to maintain life during the first year of existence. As a rule the browsing area of a particular limpet can be recognized. Removing limpets from a rock, however, quickly shows the effect of limpet browsing, for in a very short time weeds begin to grow, while areas where limpets still browse remain bare. They scrape off the film of vegetation with their radula. (See question 252.)

317. How long do limpets live? In the case of a Japanese species, the oldest specimen is reported to be seventeen years.

318. Are limpets used as food? Mexicans prize the large owl limpet as food, and it is reported to have finer meat and a more delicate flavor than the abalone. Only the foot is used, and it must be pounded in the abalone fashion before being fried in egg and flour.

319. What are keyhole limpets? They are gastropods allied to the abalones rather than to the true limpets. They are called keyhole limpets because they are limpet-like in looks and have a hole at the top of the shell.

320. How does a keyhole limpet get its hole? "When young these mollusks have a spiral shell with a marginal slit. Shelly material is added slowly until the margin below the slit is united, and then the spiral is absorbed as the hole enlarges."—(Percy Morris.)

321. What is the hole for in a keyhole limpet? To allow water which comes in near the head to escape, making an almost straight passage across the gills. The anal opening also discharges through this hole and thus contamination of the incurrent water is avoided.

322. Where are keyhole limpets found? Several species are commonly found along the Florida coast together with those of *Diodora,* mostly clinging to the sides of stones, rocks, coral and wharf pilings in shallow water. The two most outstanding forms are the Barbado chink (*Fissurella*) usually attached to rocks and pilings where the waves break against them, and the keyhole limpet (*Diodora*) which has usurped the common name for the group, and are also found attached to rocks, wharf piles and corals. On the West Coast the small, beautifully marked keyhole limpet known as the volcano shell is frequent from southern California southward. Farther north two large forms occur, the seven-inch giant keyhole (*Megathura*) which lives on the open coast from Monterey to Lower California, and the somewhat smaller species, *Diodora aspera,* which ranges along the whole coast line.

323. How do keyhole limpets feed? In much the same manner as true limpets, that is by browsing, although less is known about them.

324. Why are slipper limpets so called? Because the empty shell when seen from below looks somewhat like a slipper.

325. What are slipper limpets? Slipper limpets (*Crepidula*), while having more or less a limpet form are not really limpets but are more closely allied to the periwinkles.

Slipper limpets, one on top of another,
and two solitary ones (1 1/2")

326. Where do slipper limpets live? The slipper limpet (*Crepidula fornicata*) is common and native along the American Atlantic Coast but was introduced to the shores of England about 1880 along with American oysters. The West Coast forms are the brown slipper and the horned slipper. (See question 670.)

327. Does the slipper limpet move around? No. Since it does not need to move in order to feed, it settles firmly in one place—in fact, even more so than the true limpet. It has the peculiar habit of living in chains, one animal settling on the back of another, up to twelve or thirteen in the case of the common eastern form but up to forty in the western. Only the lowest in the series is attached to a rock shell, or some other solid object.

328. How do slipper limpets feed? Slipper limpets, while true gastropod mollusks, feed like bivalves by means of enlarged ciliated gills, drawing in a current of water and trapping the contained nutritive particles and microscopic organisms. The brown slipper limpet of the Pacific Coast "Strains from the water by means of mucous coverings of the gills. The gills are paired, one gill extending backward along each side of the foot. The animal feeds first from one side and then from the other, twisting its mouth at intervals from one side to the other to gather up the mucus with its collected food."—(G. E. and N. MacGinitie.)

329. What is the relationship between the various members of a slipper-limpet cluster? They represent successive generations of slipper limpets, the bottom one being the oldest and the top one the youngest. Each one in a series is reckoned to be one year younger than the one to which it is attached, at least in the case of the eastern form.

330. Are the members of a slipper-limpet cluster all of one sex?
Although in the case of each species of slipper limpet there is but one
kind of individual, the individual changes as it grows. The young or
small individuals (i.e., those at the upper end of the totem pole) are
all males, and like all male snails are equipped for copulation. As
growth continues however they lose their male glands and organ and
slowly transform into females, so that the lowest and largest members
of the series are mature females, while those in the middle are neither
this nor that. For persons who like the taste of words, slipper limpets
are therefore protandric hermaphrodites.

331. How do slipper limpets produce their young? In the case of
the horned slipper of the West Coast, breeding individuals carry the
eggs in the shell cavity, where they can be seen if the animal is pried
loose from its support. In the common slipper of the East Coast special care is taken of the spawn. The female constructs fifty or so mem-
branous sacs, into each of which it passes 250 fertilized eggs, tying
each sac with a short cord. The cords are then attached to the surface
(of stone or shell, etc.), on which the animal sits. After about a month
the young hatch out and enjoy a free-swimming life for a couple of
weeks. Thereafter, those that happen to settle on suitable spots—and
only a very small percentage do—grow and mature as males and then
females in typical slipper-limpet fashion.

332. Are slipper limpets always the same shape? They grow in
such a way that the shell is curved to fit the support. They are
markedly curved and somewhat narrow when growing on a curved
shell, but may be comparatively broad and flat if attached to a flat
surface.

333. Do slipper limpets do any harm? Not directly, but in England
at least they have become a serious menace to the oyster fishery
through sheer competition. Along the shores of the North Sea, where
oysters used to flourish, slipper limpets now form masses, often inches
deep, on the surface of the bottom in sheltered creeks and bays.

334. Where do moon snails live? Moon snails (species of *Polinices*)
live abundantly in the sand flats along both the Atlantic and Pacific
Coasts. At low tide, even though the mollusk itself may not be visible,

long trails like those of a mole may be seen in the sand, at the ends of which moon snails may be found.

335. Why is the moon-snail foot so large? It helps the moon snail to make a successful attack upon its victims but more importantly the great foot makes it possible for the animal to glide rapidly through soft sand in its search for food without ever breaking surface.

336. What do moon snails feed on? Mainly on clams of one sort or another, although they have no objection to another moon snail and will in fact eat almost any kind of dead flesh.

337. How do moon snails feed? A long-neck clam offers little trouble to any snail—it cannot close its shell sufficiently to keep out the moon-snail proboscis—but tightly closed quahogs and many other clams are more than a man can open without a jackknife. Moon snails have their own special equipment for the job, and they use it effectively. The attack, when it comes, is formidable, overwhelming and in effect as deadly as the spring and landing crash of the lion. The huge foot, three times as long as its shell, ripples lightly around the shell of the victim, glides over it and enfolds it firmly. Then down comes the hood, which is a protective flap on the front of the head; the long, tubular proboscis is put in place, and the drill starts to act. It is an instrument that all snails and whelks possess, a ribbon of closely set teeth known as the radula. Once the hole is made the radula continues its work by scraping the soft inner tissues into fragments and passing them back to the gizzard where the food is further ground up.

338. What animals feed on moon snails? Principally the many-rayed starfish and cannibalistic moon snails.

339. How do moon snails protect themselves? In spite of the enormous size of the crawling foot, all of it can be withdrawn inside the shell by squeezing the water from perforations around the edge of the foot. The horny lid or operculum on the back of the foot finally closes the door of the shell and the mollusk is more or less immune from attack. At the same time, so much tissue has been packed within the shell that the animal can no longer breathe and cannot live long in the closed-up position.

340. How do moon snails make their peculiar collar-shaped egg cases? "The egg cases of these great snails are, to most laymen, one of the puzzles of the intertidal world, for they look like nothing so much as discarded rubber plungers of the type plumbers use to open clogged drains. . . . Certainly there is no obvious reason for connecting the rubbery, collar-shaped egg cases with the snails that make them. The eggs are extruded from the mantle cavity in a continuous

Moon snail with egg collar
(shell 1–2 1/2"; collar 2–4")

gelatinous sheet, which, as fast as it emerges, is covered with sand cemented together with a mucous secretion. The growing case travels around the snail, taking its shape from the snail's foot as it is formed. In time the egg case crumbles, releasing a half million or so free-swimming larvae."—(E. K. Ricketts and J. Calvin.)

The egg cases are commonly known as "clergymen's collars." (See question 137.)

341. What is the abalone? The abalone is a large, limpet-like snail found in various parts of the world. There are many species, all of them belonging to the genus *Haliotis*.

Abalone (7")

342. What kinds of abalone are there? The abalone most used as food is the red abalone of middle California, which has three or four elevated openings along the edge of the shell and grows to a length of nine inches. The green abalone is smaller and has six holes, and occurs along the coast of Lower California. The black abalone tolerates more surf than its red cousin and is often found in crevices of barren rock; it ranges from Oregon to Lower California, and keeps its shell clean, unlike the red abalone which invariably carries a small forest of hydroids and algae on its back.

343. What are the holes for along the edge of an abalone shell? The holes are for the escape of water discharged by the gills, so that the animal can breathe while holding closely to the surface of a rock.

344. What happens to the holes in an abalone shell as the shell grows? More holes are added as the shell grows and the first ones are closed.

345. How do abalones hold on to rocks so tightly? The adhesion is due to fine, elongated depressions in the sole of the foot acting as suckers. The copious secretion of mucous on which the animal "slides" aids it perfecting a vacuum beneath the foot. (See question 313.)

346. How can abalones be removed from a rock? "*Haliotis* occurs most frequently on the under side of rocks and ledges, where it clings, limpet-fashion, by its great muscle foot. If taken unawares, specimens may be loosened from their support easily; but once they have taken hold, it requires the leverage of a pinch bar to dislodge them. . . . Stories of abalones holding people until the incoming tide drowns them are probably fictitious, but it is nevertheless inadvisable to try to hold them by slipping one's fingers under the shell and giving a sudden pull. We have captured them in this manner when no bar was available, but there is always danger of severe pinch."—(E. K. Ricketts and J. Calvin.)

347. How much of an abalone is eaten? Only its foot. When an abalone is to be eaten, the shell is removed and everything is cut away except the foot, which is then cut into several steaks about three eighths of an inch thick. These need to be pounded with heavy wooden mallets before the meat is tender enough to be cooked.

348. How sensitive are abalones? In *Haliotis fulgens,* the long tentacles projecting from the mantle edge are sensitive to food substances, and if a piece of seaweed touches any of them the animal whirls around to grasp it with the anterior end of the foot, which is tapering and prehensile. The seaweed is then drawn under the forward half of the foot, where it is securely held while the animal rapidly rasps and swallows it.

349. What do abalones feed on? The animal is strictly vegetarian and crawls slowly about in forests of seaweed, filling itself with large quantities of sea lettuce and kelp.

350. Are abalones legally protected? Decidedly so! Even abalone shells are prohibited from being shipped outside California and only residents and tourists in California are permitted to use abalones for food. In spite of these restrictions, however, legal-sized abalones are becoming increasingly rare and any you see along the shore are almost certain to be under the legal size for taking. Seven inches across is the minimum size permitted by law.

351. Which kinds of abalones are marketed? The red abalone is the only one marketed in North America.

352. When and how do abalones breed? The red abalone spawns between the middle of February and the first of April. It begins to breed when six years old and about four inches in diameter. In the first breeding season a female produces about 100,000 eggs, but a seven-inch specimen (probably twelve years old) will produce between two and three million eggs. A community spawns more or less at the same time, the genital products being shot out in successive clouds, the sperm white and the eggs gray-green. When an individual begins to spawn, the chemicals accompanying the eggs or sperm stimulate the other abalones in the vicinity to do likewise.

353. What is the distribution of the abalone? "Abalones are known to have occurred as long ago as the Paleozoic Age. They occur now in the Mediterranean and even on the coast of England, but it is only in the Pacific that they attain great size, the center of distribution being California, Japan and Australia. The red abalone, distinguished by its great size, the (usually) three or four open elevated apertures . . . ranges from the San Francisco area to northern Lower California, reaching its greatest development in Monterey County."—(E. K. Ricketts and J. Calvin.)

354. Do abalones produce pearls? Quite frequently. The iridescent lining of shell material will form over any small irritating particle just as in the case of pearl oysters (see question 422). One of the common irritants in the case of the abalone, however, is a minute boring clam which burrows into abalone shells like its larger rock-boring relatives. (See question 405.)

355. Are abalone shells used commercially? The colorful pearly shell is made into jewelry, but the abalone is becoming increasingly rare and neither the mollusk nor its shell can be legally taken out of the State of California.

356. What are sea hares? They are sea snails that have two pairs of tentacles on the head, which have some resemblance to the ears of a hare or jack rabbit, and whose transparent, paper-thin shell is completely hidden by folds of the mantle. They form a link between the shelled snails and the true sea slugs.

357. Where are sea hares found? The sea hare of the Pacific Coast (*Tethys californica*) extends from Monterey Bay to Lower California and lives in various environments—on fairly exposed rocky shores and on completely sheltered mud flats, in tide pools at all levels and below the low-tide line. On the east coast the West Indian sea hare (*Aplysia* species) ranges from the Florida Keys through the West Indies to the Cape Verde Islands. Usually living offshore, it comes into shallow water to breed.

358. What do the sea hares feed on? Most are exclusively vegetarian. The giant Pacific sea hare, which may weigh fifteen pounds, eats enormous quantities of seaweed. The sea hare of the Florida coast is said to feed mainly on the large seaweed codium, although the European sea hare, which is an extremely close relative, confines itself to the sea lettuce (*Ulva*). All sea hares are probably dietetic specialists though not all are vegetarian. The beautiful six-inch *Nava-*

nax, which lives among the eelgrass beds in bays south of Monterey, subsists on a diet of bubble snails.

359. How do sea hares feed? They bite off good-sized pieces of weed with their large jaws and pass them by means of the conveyor belt of radula teeth into the gut. Here they pass through three stomachs, the last two of which are lined with teeth that continue the grinding process.

360. What is the purple fluid given off by sea hares when handled or otherwise disturbed? It is similar to the purple dye of the rock purple (dog whelk) and has also been compared with the brown sepia substance produced by the octopus and squid. While its purpose in the case of the sea hare has not been determined, it probably has the same function as in the others, namely, to paralyze or offend the sense of taste or smell of an enemy. A handkerchief dipped in a mixture of seawater and pigment comes out dyed a beautiful purple, although the color washes out. (See question 261.)

361. How long do sea hares live? "Life is short; the young hatch from the strings of spawn, settle from the plankton some distance off shore and move inshore as they attain adult size and finally, at the end of a year, pass on to the shore for spawning, after which they die."— (C. M. Yonge.)

362. How many eggs can a sea hare lay? The Pacific Coast sea hare lays eggs in yellow string masses as large as two fists, each mass estimated to contain about eighty-six million eggs. A large sea hare which weighed six pounds, by no means large for this kind, laid twenty-seven such masses in an aquarium within five months, or nearly five hundred million eggs. It is therefore surprising that so few sea hares reach maturity, and this sort of observation, made by Charles Darwin along the shores of the Falkland Islands, set in his mind the seed of the theory of natural selection. (See question 352.)

363. Are sea hares of different sexes? No. Unlike the whelks and the majority of other marine snails, sea hares are hermaphrodites,— that is, contain the glands and organs of both sexes in each individual and capable of functioning simultaneously. Self-fertilization however is impossible.

364. What are sea slugs? They are sea snails that lose their shells when still microscopic and grow into slugs, usually of fantastic shape and striking colors. Many of them are among the most beautiful animals in the sea. They are nearly all flesh-eaters, but all are specialists with regard to their diet. They vary in size from about half an inch to several inches in length.

365. Where are sea slugs found? They are almost invariably found associated with the kind of food they eat. Find the food, and you can expect to find the slugs. Those known as the dorids feed, for instance, on sea squirts, sea mats and even sponges, although each species sticks to its own peculiar choice. Another group known as eolids specializes on the coelenterate animals such as the sea pansies and hydroids; again each species restricts itself to one kind or another. A few sea slugs specialize in this or that kind of seaweed.

366. What are the various processes on the backs of sea slugs? Roughly speaking, there are three kinds of sea slugs. The commonest possess a round fringe of tentacle-like respiratory gills on the back

Plumose sea slug (1–3″)
with egg mass

Sea lemon or sea slug
with egg ribbon (3/4–1″)

end, surrounding the anus. These are the dorids or sea lemons. A second group, the eolids or plumed sea slugs, have the whole upper side covered with long, simple finger-like processes, variously colored, which actually contain extensions of the liver. The third kind also has the top side covered with processes, but of a branching type. This kind is the bushy-backed sea slug known as Dendronotus, meaning "branching one," and in this case the general effect and purpose seems to be camouflage, for even large individuals several inches long are difficult to see among their natural background of seaweed and hydroids. The bright colors of the solid processes however appear to be more of a warning than a disguise. In any event the solid sea slugs seem to be unpalatable to other creatures, probably because they are loaded with the sting cells of the hydroids they have been feeding on.

MOLLUSKS—BIVALVES

367. What are bivalves? Two-shelled mollusks such as clams, oysters, scallops, mussels, etc., including the so-called shipworm.

368. How are the shells of a bivalve made? The shell is formed as the animal grows. The soft mantle, which is the outermost layer of the body, lays down successive layers of shell, each layer projecting a little beyond the last one laid down. This results in a series of concentric lines of growth marking the external surface of the shell. (See question 148.)

369. How are the two shells of a bivalve held together? By two means: by a horny ligament where the edge of the shells is thickest, which acts as a hinge, and by one or two groups of muscles which are attached to the inside surface of both shells. (See question 153.)

370. What is inside the bivalve shells? The whole body of the mollusk, consisting of a pair of large mantle folds which line the shell on the outer surface and enfold the rest, two pairs of curtain-like gill sheets, and a more or less muscular foot lying between the pairs of gills; where the foot thickens toward the hinged part of the shell lie the visceral organs such as heart, kidney and intestine.

371. Does a bivalve have a heart? Each has a heart which lies in the body close to the hinge of the shell.

372. Do bivalves have blood? They have an almost colorless blood, which does, however, have a bluish tinge. They are, in fact, true blue-blooded creatures, as distinct from those of us who have red blood, and they employ copper instead of iron in their oxygen-carrying blood pigment. (See question 509.)

373. Has a bivalve a brain? It depends on what you are willing to call a brain. A bivalve has a nerve center near the mouth, which is sensitive to touch and taste, and other centers which operate the muscles of the foot and the shell-closing muscles. The nerve center near the mouth is actually in the position of a brain, but it lacks too many features to be worthy of the name.

374. How do bivalves breathe? Water is drawn in, either through a siphon or through the gaping edges of the shell and mantle, and the oxygen it contains is taken up by the blood of the animal as the water passes over the surface of the curtain or sheetlike pairs of gills.

375. How do bivalves feed? Bivalves feed by drawing a current of water between the shells. The water is drawn in either through a siphon or through the wide, gaping edge, but is always passed out again as a relatively strong stream near the hinge. As the water passes over the surface of the gills, the minute food particles it contains, such as diatoms and other single-celled plants and animals, are sifted out and directed into the mouth at one corner of the gills. Bivalves feed in virtually no other way, and the maintenance of a steady stream of clean, food-bearing and oxygen-bearing water is essential to their existence. (See questions 90–97, 913.)

376. How large do bivalves grow? The largest of all is the giant clam of the Australian Barrier Reef and the South Pacific Islands (*Tridacna*), which may weigh more than 500 pounds, including the shell, and exceed five feet across. On the Pacific Coast, in the Puget Sound region especially, the geoduck clam may weigh as much as ten to twelve pounds, of which only a small part is shell.

377. Where do scallops live? Scallops live, for the most part, offshore in from shallow to moderately deep water, usually on gravelly bottom, but are often found a little below low-tide level wherever eelgrass beds may still be seen. Before the eelgrass disease broke up the beds in 1930, scallops were common among them and will probably return as the eelgrass recovers. (See question 132.)

378. What scallops are eaten? All are edible, but only a few are large enough to be marketable. On the Atlantic Coast the common scallop (*Pecten irradians*) is dredged by the ton by scallop fishermen, is two or three inches across, and is abundant from Cape Cod to Cape Hatteras. Farther north, ranging from Labrador to New Jersey, the giant scallop (*Pecten grandis*), five or six inches across, is also abundant; it lacks the heavy shell ribs of the other. Other species of scallop (especially *Pecten hindsii*) form the basis of an important Pacific Coast industry.

379. What part of a scallop is eaten? As a rule only the large single muscle which holds the shell together is sold and eaten, the rest of the body being thrown away, although the entire animal is edible and is said to be delicious. The smaller scallop may be eaten entire, either raw or boiled, just as an oyster, and gastronomically adventurous persons should try to obtain their scallops alive in their shells and experiment accordingly—remembering, however, that scallops have eyes and may see where they are going, whereas oysters are blind. (See question 383.)

380. Which side of the scallop lies uppermost? One shell of a scallop is always more curved than the other. This is the shell that lies beneath, with the flatter shell uppermost so that the animal is lifted

Scallop (3″), showing
eyes and tentacles

somewhat from the sea floor, thereby enabling it to take in cleaner water for food and respiration and also making it easier to take off and swim.

381. Of what use are the tentacles that are found along the mantle edge? For sensing any small creatures or particles of sand, etc., which might enter the shell unseen.

382. Can scallops move about? Scallops, except perhaps the largest, are active swimmers and have considerable control over their movements. The common scallop is very active. In the typical swimming movements the animal flaps the two shells vigorously, leaving the bottom and moving upward in a series of convulsive jerks, falling down a little between movements. The free edge of the shell goes in front so that the animal seems to be taking a series of bites out of the water. By such means they are able to migrate from place to place, and a horde of scallops swimming through the water like a school of fish is an impressive and astonishing sight.

Scallops can move in other ways. By controlling the position of the mantle curtain along the edge of the shells, the force of the water brought about by the contraction of the great muscle may be directed this way or that. If the curtain is drawn inward, the movement of the animal becomes a sudden shooting backward as the water is forced forward; or if water is forced out only at one side of the shell hinge, the animal suddenly spins around. Both movements are employed for emergency escape from enemies, whereas the rhythmic swimming action is generally used for actual travel. (See question 485.)

383. Can scallops see? The common scallop has thirty to forty brilliant blue eyes on its mantle margin just within the edges of the shells. While it has little that can be called a brain, the animal reacts quickly to shadows or to sudden movements of approaching or passing animals. If the scallop could not move, like an oyster, or was as slow as most clams, such vision would have little or no value; but scallops have a remarkable capacity for swimming and as usual sight and locomotion go together. Each little eye has a well-developed focusing lens, a receptive retina and conducting nerve fibers.

384. What are the ridges or ribs on a scallop shell for? The ribs of the two shells interlock when the shells are closed and form a much tighter and more impenetrable structure than it would otherwise be.

385. How do scallops propagate? In the common, simple manner by shedding myriads of small eggs into the water where they become fertilized by sperm from other individuals. Sexes are separate.

386. What are the enemies of scallops? The usual enemies that most bivalves have, such as starfish, crabs, carnivorous snails, and rays, except that as a result of their unusual power of movement they are much more likely to escape.

387. Where do clams live? Clams are comparatively defenseless creatures and live safely by burrowing into sand, mud, wood or rock, although only so far that they can still extend their siphons to the surface to obtain the clean water necessary for food and respiration.

388. Where are long-neck or soft-shell clams found? In shallow, muddy bottoms between the tides, with just the tips of their extended siphons exposed when covered by the tide. This species (*Mya arenaria*) also called the mud clam, is remarkable inasmuch as it thrives in brackish water, can stand temperatures below freezing and can live for many days in a medium absolutely lacking in oxygen. On the

Long neck or soft-shell clam (shell 3–5″)

Atlantic Coast it ranges from the Arctic to North Carolina. It was originally absent on the Pacific Coast but appeared there in 1880, accidentally imported with oyster spat, and is now well established in firm-soiled flats from Monterey to the Georgia Strait. In San Francisco Bay it is "farmed" by fencing off several acres of mud flats in order to keep out the skates or stingarees. On the Atlantic Coast this is no problem, and the only protection necessary is from indiscriminate clam diggers. In the Arctic the walrus feeds almost entirely upon Mya.

389. What kinds of clams are eaten? All clams are edible, but usually only those are used which are large enough and common enough to justify the effort of gathering. The principal kinds used as food are the long-neck or soft-shell clam, the hard or little-neck clam (also known as the quahog), the large surf clams, the razor or jack-knife clams, and the enormous geoduck of the Northwest Pacific Coast.

390. What parts of a clam are eaten? All of the clam is used, either whole or cut up for chowder, after it has been allowed to lie in clean sea water for several hours in order to get rid of any sand or debris it may contain.

391. How do you find clams? By hard digging in sand, mud or sandy mud exposed from half-tide down to low-tide level. In the case of hard-packed sand, small dumb-bell-shaped openings in the sand indicate the presence of a clam somewhere below. Most clams, however, are within burrows and can descend by means of their feet faster than you can dig. How far you have to dig depends on the kind of clam. In soft mud a clam rake is generally used. (See also questions concerning particular kinds of clam.)

392. How does a clam move about? By extending its flexible, muscular foot out between the shells, gripping the sand or mud and pulling the body and shells along. Some are more adept than others at doing this—for instance, the razor clams and surf clams. One clam (*Lima*) swims well on occasion, while others such as the piddock and so-called shipworm burrow actively into rock and wood respectively. (See questions 401, 404–406, 429.)

393. How do clams propagate? In the same manner as most bivalves, by shedding eggs and sperm into the water in enormous quantities. The fertilized eggs develop into small swimming larvae which disperse with the currents. The sexes are separate.

394. Can you tell the age of a clam? Most clams and other bivalves have ridges on their shells, one set more or less radiating fanwise from the hinge to the open curved edge, and the other set forming concentric lines parallel to the outer curved edge. This second set is usually unevenly spaced, so that the lines tend to bunch at intervals. Each such bunching or grouping represents the shell growth during low winter temperatures. During the spring and summer, growth is faster and the lines are farther apart. The number of the more obvious bunchings or broad bands accordingly gives you the number of years the animal has lived.

395. Why does the common long-neck or soft-shell clam have such a long neck? Because the double openings of the siphons at the top end of the neck must be able to reach the surface of the sand or mud in order to take in the life-giving stream of water and return it, while the body slowly descends into the ground as it grows. The deeper it burrows the safer it becomes and the longer the neck must be, for when danger appears the body of the clam cannot move and all it can do is to withdraw its siphon or neck.

396. What are the enemies of clams? Any clam that happens to lie at the surface of the sea floor is likely to be eaten by starfish or crabs, but their greatest enemies are the carnivorous snails which plough through the sand and mud in search of them, moon snails, and drills.

397. What are the marks or grooves seen on the inside of the clam shells? The two large more or less circular areas are where the two groups of muscles were attached that controlled the closing and opening of the shells. The long groove parallel to the curved edge of the shell indicates the line of attachment of the mantle which underlies the shell as a whole except where the muscles attach.

398. Where are geoduck clams found? Geoducks are known mainly in the Puget Sound region, but they also occur in many quiet waters as far south as southern California. Those who have tried to dig them will understand why they are rarely found and so infrequently used as food.

399. How do you catch a geoduck? Only by means of heavy labor and much persistence. "The geoduck lives in soft muck, or even fairly loose sand, for, contrary to popular belief, it is an extremely poor digger. Lest this statement arouse the ire of many people who have exhausted themselves in the fruitless effort to reach one of the animals, we hasten to explain. The geoduck lives in a semipermanent burrow that is often three feet below the surface, sending his immense siphons upward to the surface. Any disturbance in his neighbourhood causes him partially to retract his siphons, expelling contained water from them as he does so, and thus giving the impression that he is digging down to greater depths. Continued disturbance causes continued retraction, although the siphons are much too large ever to be withdrawn completely into the shell."—(E. K. Ricketts and J. Calvin.)

400. How large do geoduck clams grow? Individuals more than eight inches long are fairly common, while the record weight is about sixteen pounds, of which only a small part is shell.

401. How fast can razor or jackknife clams burrow? "A razor clam . . . depends on speed of digging for protection from wave shock. A clam that was displaced by a particularly vicious wave could certainly be reburied under several inches of sand before the next comber struck, for specimens laid on top of the sand have buried themselves completely in less than seven seconds. The foot, projected half the

Razor clam (6–9") starting to burrow

length of the shell and pointed, is pushed into the sand. Below the surface the tip expands greatly to form an anchor and the muscle, contracting, pulls the clam downwards. The movement is repeated several times in rapid succession before the clam disappears. A digger must work quickly to capture the animal before it attains depths impossible to reach."—(E. K. Ricketts and J. Calvin.)

402. How can razor clams be caught? "When the tide is in, the animals approach near enough to the surface for the short siphons to project above this, but when the water leaves the sand they retreat deeper, although their presence may be indicated by shallow depressions from which sudden jets of water and sand may be forced up by the animal. To be caught they must be approached with caution, because razor-shells are highly sensitive to vibrations and at once retreat still deeper. A sudden deep dig with a spade or fork may bring up an intact specimen, more often a broken portion, but most frequently, except to the expert, fruitless. A less exhausting mode of capture is to place a handful of salt over the hole. As this dissolves, the increased salinity irritates the animal below and it may come to the surface and project the hinder end of the shell sufficiently for this to be seized and, with a sudden jerk, pulled out intact. Any hesitation will give the foot time to take grip of the sand below and it may either succeed in pulling the shell down or else the animal may be literally torn in two between the opposing pull of hand above and foot below."—(C. M. Yonge.)

403. Why are the shells of razor clams the shape they are? The shell, and the animal as a whole, is narrow and of even width so that it can be drawn rapidly down its burrow.

404. Are cockles used as food? Rarely in America, because they are hardly abundant enough to be worth while gathering, but they have been a staple food in the coastal towns of the British Isles for centuries. There they live in the soft mud of tidal bays and estuaries, often in fantastic numbers. For instance, in a single cockle bed of

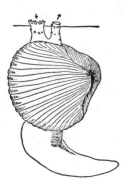

Edible cockle (3″)

some 320 acres on the Welsh coast, the cockle population exceeds four hundred and fifty million, and seemingly no amount of digging reduces their numbers for any length of time. "Cockles can move freely on the surface, and on occasion whole beds of them may apparently migrate—this is probably due to water movements. The story is told of a clergyman near Morecambe Bay who complained that half his parishioners had left him. The local industry was cockle-gathering and the animals had moved from the shores bounding his parish to others across the Bay, whither his parishioners had followed them."—(C. M. Yonge.)

405. How do the rock-boring clams work? "The shells of the boring clams, capable of penetrating coral, moderately hard rocks, and wood, are white, thin and brittle, generally elongated, and narrowed toward the posterior end, with a sharp, abrading structure on the anterior end. The borers include the wing shells, the angel's wing, and above all the piddock drills into rocks so hard that nothing short of a sledge hammer powerfully swung will break into the burrows, and it apparently drills without the aid of chemicals, using mechanical means

Rock-boring clam, or piddock (3–5″)

only. The animal twists and rocks itself on its round muscle foot (which takes the form of a suction disc that grips the rock), pressing its rough valves outward at the same time so as to bring them into contact with the walls of the burrow. Such a method of drilling is necessarily slow in ratio to the hardness of the rock, partly because the continual wearing of the shell must be compensated by growth. . . . It has caused considerable damage to concrete harbor works on this coast. . . . In some cases the animal had turned aside after penetrating the concrete jacket, in order to avoid entering wood."— (E. K. Ricketts and J. Calvin.)

406. What is a file-shell clam? The file-shell clams are found on both sides of the Atlantic, particularly on warmer waters such as Florida and the West Indies and in the Mediterranean. The file shell (*Lima*) lives in shallow waters, usually in crevices and under stones in bays and lagoons, and builds a crude sort of nest of threads of its own making, plastered with bits of seaweed and pebbles. It can also swim, somewhat like a scallop. "It has the unique power among bivalves of constructing an elaborate nest with byssus threads. Holes are left for the entrance and exit of water and the animal lies in reasonable safety within, collecting plankton from the water stream. The tissues are fringed with very many long, trailing tentacles of a beautiful red or orange color. These cannot be withdrawn within the delicate ribbed white shell, and the animal is easy prey for the first predator that encounters it outside the protection of the nest. Within the nest the tentacles are so disposed as to guide the water currents. A swimming Lima is a most charming sight. Progress is made with a series of rather languid movements with each of which the long fringe of tentacles slowly rises and then gently descends around the white shell."—(C. M. Yonge.)

407. What is an oyster? Oysters are bivalve mollusks with all the characteristic features of the group. (See questions 367–375.)

408. What kinds of oyster are there? There are many species of edible oysters, all belonging to the genus *Ostrea,* some large, some small, but all edible, including the mangrove oyster which grows in clusters on the prop roots of mangrove trees. The pearl oysters of the Indian Ocean and southwest Pacific are not true oysters but are close relatives of the common mussel. (See question 421.)

409. Where are edible oysters found? The Virginia or American oyster (*Ostrea virginica*) is extremely abundant from Massachusetts to the Gulf of Mexico and locally abundant from Maine to the Gulf of St. Lawrence. The smaller but equally delectable Olympia oyster of the Pacific Coast (*Ostrea lurida*) ranges from Alaska to middle

'Coon oysters (2") attached to mangrove roots

California. In Europe two kinds are marketed, the common British oyster (*Ostrea edulis*) and the Portuguese oyster (*Ostrea angulata*), which flourishes along the west coast of France south to Portugal. The coon or mangrove oyster, which attaches to the root props of mangrove trees, is abundant through the West Indies and Florida Keys but is less regarded as an article of food.

410. What part of an oyster is eaten? All of an oyster is eaten except the shell.

411. Can oysters be eaten at any time of the year? If they are healthy, but live oysters are more safely eaten during the colder months. They breed during the summer and are better left alone to carry out their business during those months without an R.

412. Do oysters move about? No. Once an oyster larva has settled down it never moves again except to open and close its shell.

413. How do oysters cement themselves to rocks? The oyster larva moves about over the area it has selected to settle on, testing it out with its foot. After choosing a spot for attachment, the larva circles about, wiping its large and turgid foot over the area and exuding enough cement to fix the left shell. The cement hardens in about ten minutes. As the oyster continues to grow, more cement is added from time to time by the cement gland in the foot.

414. Can oysters live out of water? For a considerable length of time, because they can keep their shells so tightly closed that no water escapes, and because their oxygen requirements are so low they can survive for a long time without being asphyxiated, particularly at low temperatures. Under refrigeration oysters can be shipped and kept alive for long periods.

415. What are the enemies of oysters? Starfish and crabs are the oysters' greatest enemies, although the flat-toothed rays also feed upon them. Apart from such obvious predators, the oyster suffers much from the attacks of the oyster drill, a small snail which drills through the shell to feed on the oyster flesh, particularly of young forms. The common slipper limpet is more of a hazard than an enemy inasmuch as it settles on the oyster shells and tends to smother the oysters.

416. Do oysters have separate sexes? As a rule oysters are definitely male or female at any one time, although not all oyster species have truly separate sexes. Virginia oysters mature first as males and begin to discharge spermatozoa when about five months old. By the time the sperm is all shed, the sex glands are already beginning to change to female and at six months are busy producing eggs. But the California oyster, having been a male and changed to a female, changes back to being a male, and from then on every few months for the rest of its life changes its sex from one to the other. All the individual oysters do this, but not all at the same time, so that in effect it works out as though the two sexes were as truly separate as mice and men.

417. How many eggs can an oyster spawn? In one season a large Virginia oyster may produce as many as five hundred million eggs and as the oyster may well live for ten years or so, it may liberate five billion during its lifetime.

418. What is oyster spat? The eggs and milt of the oyster are shed freely into the water. After about thirty hours a microscopic larva, the "straight-line larva," develops, which already has two shells. It glides through the water for eight or ten days, growing all the time, and then settles on the sea floor. It still glides along probing for a place for final attachment. This is the oyster spat—not much larger than a pin head but already shaped like an oyster.

419. Are oysters cultivated or farmed? Oysters are farmed extensively off the west coast of France where the procedure is the most elaborate but they are also farmed in Japan and on this continent in Chesapeake Bay and in Maryland estuaries. As in dirt farming, when you want to plant a crop you first prepare the ground. In the case of oysters this can be done in several ways. In public grounds, a "culch" is maintained—that is, a perpetual bed of dead oyster shells kept in good condition, so that the larvae can strike and a good spat form. This is done by dumping old shells onto the beds and in the Chesapeake alone some three million bushels of them have been planted during fifteen years. This obviously was not enough and so in Virginia there is the Cull Law, providing that all dead shells without spat must be returned to the beds, as well as all oysters less than three inches long. Only three-year-olds, between three and four inches long, can be taken. In this way an annual crop of only mature oysters is gathered. Private grounds are often cultivated in another way. Seed oysters collected from natural seeding grounds are planted and allowed to grow for three years. Then the crop is gathered and the ground cleared and replanted, as in good lumbering.

420. What are the hazards of oyster-farming? As in all farming, principally the loss of seed and the destruction of the growth stock. Chesapeake oyster farmers watch with their fingers crossed. A sudden drop in temperature could kill the larvae in a few hours, or an offshore wind might suddenly drive the surface water and larvae out to sea, and no spat will form on the waiting beds. Even when the beds are flourishing and the oysters nearly ready for market, an excessive run-off from the Susquehanna watershed can turn the upper surface of the Chesapeake so brackish that the oysters in these parts are killed. It has happened too many times. Quite apart from the elements, the oyster farmers have as much trouble with pests as the corn-belt farmers have with the corn-borer. Only instead of an insect, which cannot live in the sea, it is a sea snail. A horde of invading starfish may do more damage to an oyster bed in a short time, but this snail, the oyster drill, lives among the oysters and takes its time and toll. The toll is a high one, probably close to 50 per cent of the maturing Virginia oysters.

421. Where are pearl oysters found? Pearl oysters are not true oysters but are more closely allied to the common mussel. They are found for the most part throughout the East Indies from Ceylon to

Australia and the Malay Archipelago, and also in the Persian Gulf and along the East Coast of Africa.

422. What makes a pearl? A pearl is made of the same substance as the inner or "mother-of-pearl" lining of the shell, and both shell and pearl are produced by the living mantle tissue. Any small nucleus such as a grain of sand or the egg or larva of a parasite may serve as a center around which pearl substance may be found. Layer after layer is applied around the irritating object by the surrounding living tissue so that a round bead generally builds up, the luster of which is due to the concentric layers of the crystalline pearly substance. "The parasite which most frequently causes pearl formation in the Ceylon pearl oyster is the early stage of a tapeworm which becomes adult in a large ray which habitually feeds on these oysters, thereby infecting itself with the tapeworm. After becoming mature, the tapeworm lays eggs which pass into the sea where they may find a temporary home in an oyster. A temporary home if the oyster is swallowed by the ray shortly after, but a grave if the oyster should be provoked to protect itself from the irritating parasite by coating it with pearly substance. It may be that pearls are also formed around tiny granules of waste matter produced by the oysters themselves."—(F. S. Russell and C. M. Yonge.)

423. What are cultured pearls? They are real pearls made by pearl oysters, but they are brought about by introducing small foreign objects such as sand grains beneath the shell and mantle instead of waiting for nature to do so by accident.

424. Are pearls found in bivalves other than pearl oysters? Almost any bivalve can produce a pearl but the finest pearls are those produced by the pearl oyster. Valuable pearls however are also found in the fresh-water mussels of America, Europe and Asia.

425. Are mussels edible? The common Atlantic blue mussel, the horse mussel, and the Californian mussel are considered edible. The ribbed mussel is not. In all cases certain precautions must be taken. All mussels are dangerous if they have been collected from beds close to towns where sewage enters the sea. Bacteria are freely taken in with other suspended matter by the mussels and, although themselves un-

Mussels attached
by (1–2″) byssal threads

affected, they can become media for the spreading of pathogenic bacteria such as those of typhoid. Yet a flourishing mussel bed represents one of the densest accumulations of animal life in the world, and an acre of the best mussel ground will produce annually 40,000 pounds of mussels. Contamination is most probable in the sheltered regions where beds are most likely to be found, although on the open coast of California mussels at times ingest small marine organisms which are paralytic to humans. All in all, mussels are best left alone as human fare.

426. Where are mussels found? On both sides of the Atlantic where the seas are cool, attached to wave-beaten rocks in enormous numbers or forming extensive beds entirely covering areas of mud flats in estuaries and quiet bays, anchoring to rock, stones or to one another for mutual support.

427. What kinds of mussels are there? The commonest form is the edible or common blue mussel (*Mytilus edulis*) which lives attached to rocks of the open coast or to stones imbedded in mud flats, usually in astronomical numbers, from North Carolina northward and in Europe. The horse mussel (*Modiolus modiolus*) is a larger form four to six inches long, which lives in the crevices at the bottom of low-level tide pools along the New England Coast. Other mussels, such as the ribbed mussel, (*Modiolus demissus*), are more abundant farther south, while the California mussel (*Mytilus Californicus*) extends along the Pacific Coast from Alaska to Mexico in much the same way as the common mussel of the Atlantic, equally at home on surf-swept rocky coast.

428. How do mussels attach themselves to rocks? "Mussels are the commonest of bivalves that are attached by a byssus. The threads of this are produced by a gland within the substance of the foot and they issue from it in the form of a thick fluid that runs along a groove which extends down the hinder surface of the thin and very extensive foot. This organ plants the threads. It is directed just in one direction, then in another, the sticky fluid running along the groove and spreading out into a rounded disc of firm attachment where it meets the rock. Almost immediately each hardens into a tough thread. Finally the mussel is attached by a diverging mass of threads like the guy ropes of a tent."—(C. M. Yonge.)

429. Can mussels move about? When it is desirable, mussels can change their position, although very slowly and for very short distances. If, for instance, they are attached by their byssus threads to stones below the surface of mud, and more mud descends upon them, they can cast off the old attachment and make new and longer threads that enable them to remain on the surface. Mussels attached to rocks can break old threads and attach new ones in such a way that they can draw themselves into new positions without ever actually letting go, on the same principle as a man climbing a rocky cliff. The threads are tough and often golden brown, and are very similar to those of the fan shell which were used in older times to make the famous cloth of gold. (See question 446.)

430. How do mussels resist the action of the waves? "In whatever direction the sea strikes, some of the threads are in a position to take the strain. The shape of the mussel helps in resisting wave action. The threads are directed forward so that the animal tends to swing round with the narrow end facing the force of the sea, although in a dense mussel bed the animals are so tightly packed that movement is impossible. If the threads are broken new ones are readily formed and the animal may move about for a short time by means of the foot before it attaches itself."—(C. M. Yonge.)

431. How fast do mussels grow? Mussels can establish themselves more quickly and in denser masses than any other shore animals. At the end of World War II, when the Germans were driven from the Netherlands, the dykes guarding the island of Walcheren were broken and the sea flooded the greater part of the island. A year later, when the dykes had been closed again and the water pumped out, the surface of roads and the sides of houses and fences were found covered with mussels. Mussels even hung like bunches of fruit from the branches of trees.

432. What are shipworms? Shipworms are not worms at all but are bivalve mollusks, mostly species of Teredo, which bore into wood and have a worm-shaped body. They are eaten as delicacies by Australian aborigines and sometimes by an odd naturalist.

433. How much damage do shipworms do? "Most notorious of wood-borers is the shipworm, *'calamitas navium'* of Linnaeus, known and dreaded since classical times when it riddled the planking of Greek triremes and Roman galleys. Later it was to destroy Drake's Golden Hind and, in 1730, to threaten the very existence of Holland by attacking the dykes. It still remains a menace to unprotected boats and to pier piles; between 1914 and 1920 its sudden spread through San Francisco Bay caused damage estimated at some ten million dollars."—(C. M. Yonge.) During the San Francisco attack, ferry ships collapsed and warehouses and loaded freight cars were pitched into the bay, and all piling destroyed.

434. How fast can shipworms destroy pilings? A heavy attack will reduce a new, untreated pile to the collapsing point in six months, and survival for more than a year is unlikely. The rate of boring is more rapid in comparatively young animals; a three-month-old individual bores about three quarters of an inch a day.

435. Do shipworms feed on wood? At least to some extent, although they also take in water and its contained nutrient organisms through

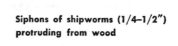

Siphons of shipworms (1/4–1/2")
protruding from wood

their siphons. All wood fragments scraped off by the shells in boring are passed into the mouth and through the length of the gut before reaching the exterior, and during this passage at least some of the cellulose in the wood is digested and converted into sugar.

436. How do shipworms burrow into wood? The mollusk rasps the wood with its pair of somewhat wing-shaped shells, rocking on its foot eight to twelve times per minute in much the same manner as in the rock-boring clams. The cutting organ, consisting of the foot muscle and the two shells, rotates so that a perfectly cylindrical burrow is cut.

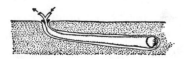

Shipworm (3–5″) in burrow
with boring shell at blind end

There are two distinct grades of rasping surface on the shell, and the animal's intestinal contents show two correspondingly different sizes of wood particles. As the animal bores, the body elongates, for the shipworm is attached to the burrow at the hinder end of the body, where a pair of paddle-like limy plates or pallets are present in order to close the opening when necessary. (See question 405.)

437. Do shipworms' burrows ever meet? No. The animal seems to be aware of the nearness of another burrow and turns aside to avoid it.

438. Do the burrows of shipworms follow a definite course? In newly infested wood the burrow runs along the grain, following the course of least resistance, but if the infestation is intense the wood becomes riddled with burrows passing in every direction.

439. How do shipworms infest wood? Small living larvae are set free during the breeding season by the million from the tubes in infested wood. Each larva is already equipped with a minute bivalve shell. If water movements carry it by chance against a wooden surface, there it remains. Once in contact with wood, the larva begins to change. A small foot appears, and a temporary attachment is made with a single byssal thread like that of a mussel. The shell then changes in shape from a protective covering into a most efficient cutting tool, and burrowing begins.

440. What happens when shipworms have completely riddled a piece of wood? If the wood is part of a weight-bearing structure it collapses suddenly. Shipworms, however, if unable to continue burrowing in any direction, proceed to lay down a thin layer of lime over the internal surface of the burrow and continue to live and reproduce for some time.

441. How can submerged wood, whether pilings or ship bottoms, be protected from shipworms? "The life of a pile may be prolonged to three or four years by a chemical treatment, such as creosoting, but nothing keeps the animals out for long. Copper sheathing, which did good service in protecting ships for so many years, is impractical on piles because of its expense, the likelihood of being stolen, and the ease with which it is damaged by contact with boats and driftwood— also the sheathing cannot extend below the vulnerable mud line, and the mud line is likely to be lowered by eroding currents. Numerous other jacketings have been tried, but not even concrete jackets are entirely satisfactory, for when the wood borers are thus thwarted the concrete borers come into their existence."—(E. K. Ricketts and J. Calvin.)

442. Can shipworms live out of water? Shipworms cannot live out of their burrows in wood, but when their siphons are withdrawn a pair of pallets or limy plates close the openings so that water is retained within the burrows. In this way the mollusks are able to survive even after the wood has been out of water for several weeks.

443. What is the sex life of a shipworm? Every individual shipworm matures first as a male. After shedding its sperm it continues to grow and matures again later as a female. It does not, therefore, fertilize its own eggs but only those of older individuals. Fertilization occurs within the burrow of the female, by sperm drawn in with water through her siphon, and swimming larvae are liberated ready to infest more wood in the same general territory in which it is born. The ovaries of a single shipworm may contain more than one million eggs at one time. (See question 416.)

444. Where do jingle shells come from? The attractive jingle shells (*Anomia*) tossed high along the beaches of the whole Atlantic seaboard, and highly prized by children, are thrown there so abundantly by the tidal surf because of their lightness. The animal itself is common below low tide throughout the shallow waters, fastened to oysters and stones.

445. Are jingle shells bivalves or univalves? Jingle-shell mollusks are typical bivalves, but the shells found on the beach are only the top shells of the animal. The lower shell is molded and cemented so firmly to whatever it is fastened to that it never breaks free and only the top shells are washed ashore.

446. Where are fan shells (or pen shells) found? The fan or pen shells (species of *Pinna*) occur from North Carolina to the West Indies and are also especially common in the Mediterranean. They grow to a length of from six to nine inches. The mollusk lives buried in gravelly mud or sand with its pointed shell down and the open, sharp edge projecting about one inch above the surface, a hazard to bare feet at low tide. The lower end is anchored to stones by means of byssus threads similar to those of mussels, to which the animal is related. The threads, however, are longer and are greenish-gold in color. They have been long in use commercially in the Mediterranean, where they were woven to make the cloth of gold worn by Roman emperors and others. The industry has never quite died out, and small garments are still made at Taranto in Southern Italy and sold as souvenirs.

"The pen shells begin to appear in numbers on the beaches below Cape Hatteras and Lookout, but perhaps they, too, live in the most prodigious numbers on the Gulf of Florida. I have seen truckloads of them on the beach at Sanibel even in calm winter weather. In a violent tropical hurricane the destruction of this light-shelled mollusk is almost incredible. Sanibel Island presents about fifteen miles of beach to the Gulf of Mexico. On this strand, it has been estimated, about a million pen shells have been hurled by a single storm, having been torn loose by waves reaching down to bottoms lying as deep as thirty feet. The fragile shells of the pens are ground together in the buffeting of storm surf; many are broken, but even those not so destroyed have no way of returning to the sea, and so are doomed."—(Rachel Carson.)

MOLLUSKS—CHITONS

447. Where do chitons live? They are found firmly attached to rocks on open coasts. The majority live under rocks or ledges during the day and are found on the exposed sides only at night. Chitons generally

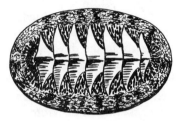

Rock chiton (1/2–6″), with eight shells and leathery margin

occupy rock depressions which give them a better chance to resist the force of breaking waves. Sometimes in the spring great congregations of the giant *Cryptochiton* on the west coast gather on rocky beaches, having presumably come in from deeper water to spawn.

448. Are chitons called by any other names? They are often called "sea cradles" because of the way they curl up when detached from the rock. In some localities they are referred to as "butterfly fish," a name suggested by the shape of the separate shells; while in the West Indies natives call them "sea beefs." The giant red *Cryptochiton* of the Pacific Coast is sometimes called the "gum boot."

449. What is the advantage of the peculiar shell? "The eight over-lapping shell plates are articulated so that the oval body can be bent readily and the long foot cling firmly by conforming to the outlines of the most uneven surface. The plates are imbedded in a tough tissue which expands around them so that perfect contact with the rock can be maintained when the tide is out."—(C. M. Yonge.)

450. How can chitons be removed from a rock? Sometimes by a sudden sideways blow delivered before the animal has had a moment to pull itself as tightly as possible against the rock, or else by slipping a thin blade between the rock and the animal's muscular foot, which destroys the suction.

451. What holds a chiton so firmly to a rock? The large muscular foot with which they crawl can be clamped down tightly as a sucker, so that in the case of large individuals you may require a chisel to pry it loose. (See question 313.)

452. Do chitons move about? Chitons are able to glide about on rocky surfaces by means of their muscular foot. With the exception of the black West Coast *Katherina* which remains on the exposed surface of large rocks, chitons generally retreat beneath rocks and ledges during daylight and wander about to feed mainly at night. They are reported to return to their original position early in the day, possibly being guided by the mucous secretions left on the outgoing trail. Not all chitons move around and "home" in this way. One West Coast form (*Nuttallina*) never leaves its adopted place; some rock depressions are thought to have been inhabited by generations of *Nuttallina* for thousands of years.

453. What do chitons live on? Chitons feed in the same manner as the various kinds of sea snails, scraping off their food by means of a file-like ribbon or radula located within the mouth. Most kinds live on the minute algae and diatoms which form a surface film on rocks, although some eat seaweed debris.

454. How large do chitons grow? Along the Atlantic Coast the common kind of chiton (*Chaetopleura*) rarely exceeds an inch in length. More than 100 species occur along the Pacific Coast; of these the common large black chiton (*Katherina*) grows several inches long, while a brick-red species of *Cryptochiton* which ranges from southern California to Alaska and Japan is the largest in the world and may reach a length of thirteen inches. Other West Coast forms (*Mopalia and Stenoplax*) are about four inches long which is also the average size of the so-called "sea beef" used as food in the West Indies.

455. Are chitons edible? The larger kinds are edible, but tough unless well beaten. Only the large muscular foot which attaches the animal to the rock is eaten. They are a regular article of food in the West Indies, where the natives pry them up with their knives and cook them, under the name of "sea beef." Ricketts and Calvin, however, say of the giant Pacific Coast chiton (*Cryptochiton*): "It is reputed to have been used for food by the Coast Indians, and was eaten by Russian settlers in southeastern Alaska. After one experiment the writers decided to reserve the animal for times of famine; one tough, paper-thin steak was all that could be obtained from a large *Cryptochiton,* and it radiated such a penetrating fishy odor that it was discarded before it reached the frying pan."

456. Can chitons defend themselves? Against wave action and sea creatures generally, by simply holding so tightly to the rocks that they cannot be dislodged. If by accident they do become detached, they curl up in much the same manner as the pill bug or wood louse, so that only the outer armor of shell is exposed.

MOLLUSKS—TUSK SHELLS

457. What kind of mollusk makes the tusk (or tooth) shell? Tusk shells (*Cistenides*) are fairly common on the Atlantic Coast from Cape Hatteras north and also along the Pacific Coast. They are unique in having openings at both ends. The mollusk is put in a

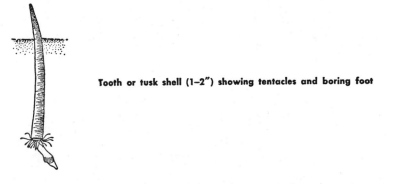

Tooth or tusk shell (1–2") showing tentacles and boring foot

category by itself, being neither a gastropod, bivalve nor chiton, and has the class name Scaphopoda more or less to itself. When alive a foot or fringe of tentacles protrudes from the wider end of the shell and digs for food in sand or sandy mud, the animal living in an oblique position, head downward, and using the smaller shell opening at the top for incoming and outgoing water. West Coast Indians strung the shells on deer sinews and used them for both trade and ornament, dredging for the live mollusks from canoes by means of long rakes.

MOLLUSKS—OCTOPUS AND SQUID

458. Can you find octopi along the shore? Small octopi, especially young individuals, may be found along the coast of Florida and southern California, in pools or among the roots of turtle grass in

Octopus (1–3')

shallow water. From Santa Barbara southward on the Californian coast a small species is common under the rocks at low tide, while south of Los Angeles octopi have moved into estuaries where there are suitable pools containing rocks arched over the mud.

459. Where do octopi live? They live in holes between or under the edges of rocks, in old shell beds, or in the crevices of reefs, for they are timid and retiring creatures.

460. How can skin divers locate octopus lairs? By examining the sea floor at the base of rocks and reefs for telltale piles of empty bivalve and crustacean shells discarded by the animals after feeding.

461. Are octopi dangerous? Not in the usual sense of the word. A carelessly handled octopus can give you a nasty nip, while a frightened octopus that gets tangled up with a swimmer under the impression that he is something secure to hold on to might conceivably cause him to drown. Both contingencies are readily avoided. Fishermen along the coasts of Europe who take them for bait or for food usually stick their thumbs into the mantle cavity and practically turn the octopus inside out, thereby rendering it helpless while yet alive.

462. Do octopi do any harm? None to speak of, except when they migrate in large numbers to an area usually as a result of changes in sea temperature, when they may do much damage to crab and lobster fisheries. This, however, is a hazard for fisheries in the English Channel rather than one affecting the coasts of North America.

463. How many arms has an octopus? Eight and only eight, never more and never less—in contrast to its relative the squid, which has ten.

464. Has an octopus any hard parts? Only its horny beak, which is like that of the parrot. Consequently, an octopus can squeeze through remarkably narrow cracks without hurting itself, or it can draw back into crevices difficult for any other largish creature to enter.

465. Can octopi swim? They only swim when they must, or if on migration. Then they swim backwards in a series of jerks employing the same method of jet propulsion used by the squid. (See question 485.)

466. What kind of eyes do octopi have? Although an octopus is as truly a mollusk as a snail or an oyster, it has evolved a pair of eyes comparable in every way to our own, complete with cornea, iris diaphragm, compound lens, focusing apparatus, an elaborate retina as a recording screen and a darkened containing wall. In other words, the octopus eye is a camera eye like that of all the backboned animals, and it is remarkable that two such basically different groups of animals should have independently gained the same complex kind of visual apparatus.

467. What senses other than sight do octopi have? The usual senses that snails possess, but very highly developed. The tentacles are extremely sensitive to touch, the mouth region to taste, the gill chamber to the chemical quality of sea water. Gravity-sensitive organs are present in the brain.

468. How good is an octopus brain? For its size the octopus brain is astonishingly good. It is at least as good as those of fish, and probably better than most. Outside of backboned animals it is the only creature that appears to possess true intelligence. Experiments conducted at the famous aquarium at Naples, Italy, have shown that part of its brain stores up memories and that the animal has a marked ability to learn from experience.

469. How do octopi change color? Pigment cells of several colors are scattered through the skin of the body and tentacles. The cells can expand into comparatively broad sheets or contract to invisible points, under stimulus of innumerable fine nerves controlled from the brain. If all contract, the animal goes pale all over. If all expand, it becomes dark brown or purple. If some kinds expand but not others, various intermediate colors and patterns appear. The changes often follow one another in a startling way as long as the creature is disturbed. When it finally settles down, it tends to take on the general tone of its surroundings so as to camouflage its presence.

470. How do octopi use their tentacles? They are used for a variety of purposes: for obtaining food, for fighting and mating, for locomotion, and for feeding. All eight tentacles bear several rows of very efficient muscular suckers so that the length of each arm or tentacle has grasping power. The suckers are operated by nerves from the

Octopus swimming and walking

brain, and the equipment as a whole has a capacity for manipulation exceeded only by that of the human hand. "One that we had we fed by firmly grasping in a closed hand a piece of fish or clam meat and then putting the hand into the water. The octopus would at once take hold of the hand and insert its tentacles between the fingers so that it could pull out the food."—(G. E. and N. MacGinitie.)

471. How do octopi feed? By the combined use of tentacles and beak, the tentacles holding the victim and the beak used to open its shell. Thereafter the octopus, which has such a narrow gullet it can swallow only liquid or highly macerated food, relies mainly upon the great power of its digestive juices. These are poured out of the mouth and into the shell of the food animal, more or less digesting it within its own skin and then sucking up the predigested mush. According to the MacGinities, "When an octopus captures a crab or other crustacean it envelopes the animal with its skirt, or web of tissue between the bases of the tentacles. Then it secretes a poisonous substance from its mouth which very quickly kills the crustacean. It then opens the crustacean at the junction of the back and abdomen, just where the skeleton would naturally open for shedding, and within half an hour removes every particle of flesh from inside the skeleton, right down to the tips of the legs. The skeleton may be torn apart at the joints, but the solid parts are intact, not crushed."

472. What do octopi eat? Mainly bivalves such as mussels, scallops and clams, since these have no chance to escape; secondarily crabs, crawfish and lobsters; and if possible fish, although a fish must come close enough for an octopus to reach out a tentacle to get it, for otherwise fish are much too fast for an octopus to chase. If a crab or a partially disabled fish should pass close to the opening of an octopus lair, the tentacle snaps out like a whip to capture it. Different species, however, have their own particular preferences. The Northern Californian octopus (*Octopus apollyon*) feeds mostly on crabs, the southern form (*Octopus bimaculatus*) mainly on bivalves. The East Coast rough-backed octopus (*Octopus rugosa*) on both crabs and bivalves. Young octopi of most species feed on hermit crabs.

473. How do octopi breathe? By means of gills in the mantle cavity, just as in the case of other mollusks, except that the blood-circulatory system is more efficient, being maintained by three hearts in place of one, and that the water circulation is also extremely effective. Water is drawn in through a wide but slit-like opening around the collar to circulate over the gills and is passed out through the prominent funnel.

474. What enemies do octopi have? They are preyed on principally by the moray eels, those vicious marine relatives of the common eel.

The moray goes down the octopus lairs like a snake after gophers, and makes a meal of any octopus it finds at the end. (See question 990.)

475. What happens if the tentacle of an octopus is torn off? As long as the animal remains healthy, a new tentacle appears in place of the old and eventually grows to be the same size as the rest.

476. How long do octopi live? Little is known except that the smaller kinds probably only live for only two or three years, and that the larger ones live probably for many years. The only records are of aquarium animals, for there is no means of determining the age of an octopus merely by studying its anatomy.

477. Can an octopus live out of water? Not for very long. While fresh it can move actively on the ground out of water, but lack of water in the mantle or gill cavity and rapid loss of water through its skin soon result in death.

478. How large do octopi grow? As a rule octopi do not exceed three to four feet from tip of body to tips of tentacles. Legend has exaggerated the size and in any case has undoubtedly confused octopus with squid. Pacific Coast octopi are known, however, to weigh as much as 110 pounds, and when spread out to be able to touch the edge of a circle ten feet across. Larger ones do exist but have not been recorded. Most shallow-water species are comparatively small.

479. Does an octopus have a shell? None whatever, although all its closer relatives have at least a trace of one.

480. How does an octopus defend itself? Through a combined use of its ink and its intelligent agility and awareness. The ink is discharged through the funnel or siphon and spreads out in the water like a smoke screen. Yet the effect of the ink seems to be more chemical than visible. "When an octopus discharges its ink in the face of an eel, the latter continues to hunt for the octopus long after the 'smoke screen' has disappeared. But the sense of smell of the eel has been so paralyzed that we have often seen the moray actually put its nose against the octopus and not know the octopus was there. Depending on the 'dose' it has received, from one to two hours later the moray can again scent the octopus and proceeds to attack it again. . . . The octopus may discharge with a second or third time, but the amount becomes less with each discharge."—(G. E. and N. MacGinitie.)

481. How do octopi breed? At breeding time one arm of the male becomes enlarged and modified as a copulatory organ. He charges this arm with a packet of spermatozoa from his generative orifice and deposits the packet within the mantle of the female. The female later, in the course of a week or so, deposits about forty festoons of eggs on the underside of her rocky lair. The eggs are nearly transparent, and the festoons resemble long bunches of miniature sultana grapes. Meanwhile she assumes a brooding position and gently waves her arms over the clusters so that water circulates among the eggs, watching over them for about two months, during which time she eats no food. Eventually the embryos hatch out as miniature octopi and go about their own business. A female northern octopus of the Pacific Coast weighing five pounds laid about 45,000 rather small eggs which took six weeks to hatch, but a half-pound southern octopus of the same coast laid about 600 eggs of a much larger size, half an inch long, which took four months to hatch. The mother ate no food for all of this time and cared for her eggs in the usual manner.

482. Is any use made of octopi? They are dried and used for food both in the Mediterranean and throughout the Orient. They are also used as bait by fishermen. Persons who should know say they taste better than oysters and as well as most clams. Southern Californians are beginning to acquire a taste for them.

483. What are the relatives of the octopus? Apart from mollusks in general, the octopus belongs to a group known as the cephalopods or head-footed creatures, which includes the various octopi, the squids, the cuttlefish, the curly and the paper argonauts, and the somewhat mysterious owner of the ram's-horn shell (*Spirula*) frequently found on Southern beaches.

484. Where can squid be seen? They are often to be seen darting about in shallow water early or late in the day, particularly during the spring months when they come close to shore to deposit their eggs.

They are seen to best advantage, however, in large marine aquaria, where their beauty of form, color and movement becomes evident. You may be fortunate in seeing them in large numbers when they follow schools of fish into coves and bays.

485. How does a squid swim? It can do everything a fish can do, but it does so in its own way with its own unique equipment. It is streamlined for speeding backward, not forward like a fish, and is reminiscent of that mythical backward-flying bird that doesn't care where it's going but would like to see where it's been. It has side fins for stability and planing up and down, like the pectorals of a fish. But where a

Common squid (8–12″)

fish sculls with a driving thrust of muscle and blade against water, the squid draws water quickly in through the wide mantle opening and then sends it like the jet of a rocket out through the funnel. It is truly jet propulsion, based on hydraulics. There is a little more to it than this, for by directing the opening of the funnel forward or backward the jet can be made to propel the animal backward or forward as the case may be.

486. How many tentacles has a squid? Ten, of which two are generally longer than the rest. The longer tentacles are used mainly more for capturing food and for mating purposes, the shorter ones for handling food already captured.

487. How do squids change color? By the same means as the octopi and apparently for the same general purpose of making themselves as invisible as possible—dark brown when close to the sea floor, pale when swimming near the surface at night.

488. What kind of eyes do squids have? Similar to the eyes of the octopus. (See question 466.)

489. What other senses do they have? The same as the octopus. (See question 467.)

490. How good is the brain of a squid? Probably as good as the octopus brain (see question 468), although less is known about it.

491. How do squids feed? In much the same manner as the octopi except that as a rule they must eat and swim at the same time. (See question 471.)

492. What do squids feed on? Principally on schools of fish, small squids feeding on small fish and large squids on larger kinds.

493. What happens if the tentacle of a squid is torn off? As in the case of the octopus a new tentacle regenerates, a likely event in both animals since each kind is attacked and eaten by other creatures, the octopus by the moray eel and squid by the sea mammals especially. In one of the smallest squids (*Rossia*) the pair of long tentacles used in mating are always torn off and left behind and a new pair grown.

494. Do squids have shells? They have a transparent, light horny shell completely hidden in a deep pocket passing down the back from the region of the collar.

495. What purpose does the squid shell serve? To stiffen the body. It is no longer used for protection, but like the backbone of a fish, that is, as a necessary stiffening and supporting rod for a fast, streamlined swimmer.

496. How do squids breathe? By the same means as octopi. (See question 473.)

497. What enemies do squids have? Mainly the sea mammals—that is, the whales, porpoises and seals, and certain of the larger and faster fish.

498. How do squids defend themselves? They have ink sacs like those of octopi and expel the ink into the water when alarmed. The ink confuses their enemies' senses of smell and sight. (See question 480.)

499. How large do squids grow? The common kinds along the coasts and visiting the bays are small species from one to two feet long (*Loligo*). The so-called giant squids of the Pacific Coast run about ten feet over all, with six-foot tentacles and a four-foot body. But the real giants are those of the deep and open ocean, where the great sperm whales feed on them. The largest of these so far recorded was about sixty-six feet long and would have weighed about forty-two tons.

500. How do squids breed? In virtually the same way as octopi (see question 481) except that the female does not take care of her eggs after she has attached them to rocks. The egg masses, however, are rarely tampered with, since they appear to be unpalatable or indigestible to the crabs and fish that are in a position to eat them. "With the exception of the annelid *Capitella ovincula, Patiria* (a sea urchin) is the only animal we know of that can digest a string of squid eggs, and a starfish required seventy-two hours to do so. . . . We have fed squid eggs to sea anemones, only to have the mass regurgitated unharmed an hour or two later, after which the squid embryos continued their development none the worse for their adventure."—(G. E. and N. MacGinitie.)

501. Can squid eggs be found along the shore? The egg masses are usually found fastened to the sides of rocks in sheltered regions just below the level of extreme low tide. Each mass consists of slender white opalescent fingers about two inches long, each containing about 100 eggs. Forty or fifty fingers are generally clustered together and are probably the product of more than one squid. If by chance you find squid eggs on the point of hatching and can watch the newly hatched squid in a glass of water with a hand lens, you will see one of the most enchanting sights in this world.

502. What use is made of squids? They are dried and prepared as food in the same manner as the octopus and are eaten by the same peoples of the Middle and Far East. In China especially, however, the ink from the ink sac is collected and sold as the famous India Ink, although modern synthetic substitutes are now displacing it.

503. Where do "cuttle bones" come from? They are the light but limy internal shells of the common cuttlefish (*Sepia*) of European and Mediterranean shores. The cuttlefish has ten tentacles as in the case of the squid, to which it is closely related, and searches for crabs and mollusks among the inshore beds of seaweeds. Unfortunately it does not enter our territory. The "bones" or shells are well known since they are imported for the use of caged birds.

504. What kind of a mollusk lives in the ram's-horn shell? The ram's-horn shell (*Spirula*) is occasionally thrown up on beaches from Cape Cod to the Gulf of Mexico. The animal that secretes it lives in deep water in the Caribbean region and elsewhere, associated with rocky territory. It is a small, red, brown-spotted animal, related to the squid, with eight short tentacles and two long ones, altogether about three inches long, and has relatively large eyes. The shell is peculiar, consisting of a series of separate chambers of increasing size.

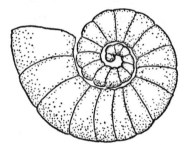

Ramshorn shell (1″)

The animal occupies only the last and largest, although the greater part of the creature extends far out beyond the shell limits. When it was small it occupied a minute chamber at the inner end of the spiral. As it grew it moved out and secreted a new seat and wall around it, doing so at intervals throughout growth until about thirty successive end walls have been produced. The animal then dies and its shell floats and drifts with the currents to end up on some beach where a curious human may notice it and wonder at the tiny shell so like a ram's horn.

ARTHROPODS — CRUSTACEA

505. What kinds of crabs are commonly seen out of water on rocky shores? The green crab (*Carcinides*) is the common shore crab along the Atlantic Coast from the Chesapeake to Nova Scotia, and are to be found beneath rock weed and boulders in the lower tidal regions

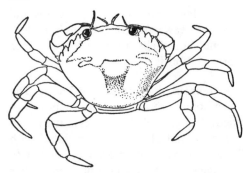

Shore crab (young green crab)
(1–4")

and on the shallow water mud flats farther down. The red cancer crabs
are usually seen along the New England Coast on the sides of rocky
ledges a little below low tide, but may at times be found in spray-
splashed crevices a few feet above the water. On the Pacific Coast the
pugnacious little square-shelled rock crabs (*Pachygrapsus*), dark red
or green, scramble over the rocks, particularly at night, from Oregon
to the Gulf of California. The purple shore crab (*Hemigrapsus*) is
its runner-up. (See questions 507, 511, 535, 537.)

506. How well do crabs see? Crabs have two well-developed eyes of
a kind borne at the ends of movable stalks so that when the stalks are
raised from their protective grooves a crab can see more or less well
in every direction. The eyes form probably only a very crude image
but are exceptionally good for determining both movement and location of an object which is either moving or has suddenly stopped moving. We have seen a ghost crab at night under a good light instantaneously seize a fly which landed within range. No performance of its kind
could have been faster or more accurate.

507. What do crabs eat? Most of them will eat any kind of flesh they
can get their claws into. The common green crab of the Atlantic is one
of the most voracious creatures along the shore and also one of the
most pugnacious (in France the name for the animal is "the angry
crab"). Some kinds of crabs are more discriminating in their diet than
others, but the common square-shelled rock crab of the Pacific Coast
feeds on anything from dead or incapacitated fish to the film of vegetation which covers the rocks.

508. What are all the peculiar appendages on the front side of a crab below the eyes? There are six pairs of appendages, each pair with its own particular shape and action, which work together to bring a stream of water into the mouth region and at the same time to handle and cut up whatever the crab may be feeding on, for the gullet of a crab is very small and only finely divided food can be taken in.

509. How do crabs breathe? Crabs have an exceedingly well-developed set of stiff, closely branched gills on either side of the body where the legs join the shell, although the shell must be broken to see them. Anyone who opens a crab to extract the meat cannot avoid noticing them, for they occupy a great part of what seems to be the interior. In life, however, a constant stream of water is passing over them, and each gill is, in fact, an upper branch of one of the legs. Blood from the heart is circulated through the gills. The blood is bluish and has the same copper as in most mollusks (see question 372).

510. What enemies do crabs have? They are hunted along the shore and in shallow water by gulls and herons, while in the sea itself many fall victim to starfish, octopi and certain fish with teeth strong enough to crush them.

511. Do crabs live along the shore all year round? True shore crabs like the green crab and the fiddlers remain on or close to shore throughout the year, but others such as the blue crab and the cancer crabs retreat to deeper offshore waters during winter months and for breeding purposes. They come inshore during spring, and the young forms use the shallow water and shore pools more as a nursery. (See questions 505, 516, 535, 538, 551, 597.)

512. Do crabs have tails? Every true crab has a tail which it cannot use for swimming. It is always firmly tucked forward beneath the body like the tail of a frightened dog. When a crab is examined from below, the tail is easily recognized by its more or less triangular shape and its division into transverse sections or segments. It can be pulled away from the body surface.

513. Is a crab's tail used for any purpose? Only for holding or carrying the mass of developing eggs during the breeding season.

514. Why do crabs run sideways? There are two possible answers, one mechanical and one purposeful. The mechanical reason is that the limbs and muscles of a crab are so arranged that any other locomotory direction is very difficult. The value reason is that a creature that runs fast from side to side is much more likely to escape capture than one that simply runs away in the line of pursuit. Any person who has tried to catch an actively running crab on rocks or sand will recognize the truth of this.

515. Do all crabs run or move sidewise? Yes. All true crabs do— that is, all crabs in which the tail is always tucked forward and never used for swimming. Spider crabs, however, do manage to walk obliquely as well as sideways.

516. Can crabs swim? Some can, some cannot, and some do better than others. True swimming crabs occur along both the Atlantic and Pacific Coasts. Both the blue and the lady crab are swimming crabs

Swimming crab (blue crab) (5–6"); note paddle-like hind legs

and are notable for the paddle-like shape of the last pair of legs, which form a pair of oars. (See questions 538, 539.) Shore and bottom-dwelling crabs have hind legs with pointed tips for getting a grip on rock and sand.

517. Can crabs replace lost claws or legs? Yes, although it takes considerable time, particularly in a full grown crab—a matter of months at least or even years, according to age and size of the crab. (See questions 518, 519, 545.)

518. How are the limbs of crabs replaced? Replacement or regeneration is limited by the nature of the shell in the same way as growth generally is limited. The first sign of a new claw or a new leg is a small knob which can be seen after the first molt or shedding of the shell after the loss has occurred. After the next molt a small limb is recognizable, and thereafter with each molt the limb becomes larger, until finally, if the animal itself lives long enough, it is the same size as its opposite member on the other side of the body.

519. If a crab's leg or claw is broken off, why doesn't it bleed to death? For two reasons; the almost colorless blood coagulates quickly while a crab leg, when injured, is snapped off at the base, no matter where the injury occurs, the snapping-off being performed automatically by special muscles which also more or less close the wound.

520. What kind of blood does a crab have? A very pale bluish blood in common with all the crustaceans and most mollusks, in which the respiratory pigment has a copper basis in place of the iron of the red hemoglobin of other creatures.

521. Do crabs lay eggs? Eggs are fertilized before they leave the body of the female. As they are shed they become attached by fast-

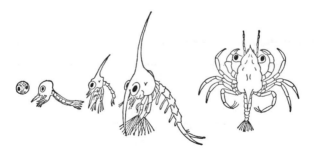

Free-swimming stages of common crabs (1/50–1/10"), from embryo in egg to last stage before descending to sea bottom

hardening glue to the small appendages of the tail, so that the large mass of eggs forces the tail away from the body. The tail in fact serves primarily as a cradle.

522. How long do crabs live? In a general way, the smaller the crab the shorter its life. The smallest kinds live probably for only two or three years. The edible blue crab lives for five or six years, while the big cancer crab of the Pacific fish markets has an average life span of eight years, with a maximum of ten.

523. How large can crabs grow? The largest crab is a spider crab of the North Pacific which lives in comparatively deep water from the coast of Japan to Alaska. When stretched out it may measure ten feet from the tip of one claw to the tip of the other, the body itself being about twelve by eighteen inches across. Large rock crabs species of *Cancer*), found along the northern Pacific coast of America and also in northern European waters may grow to be fifteen inches wide, with massive, chunky claws in proportion.

524. What is a soft-shell crab? A crab which has very recently shed its shell and has not had time for the new one to harden.

525. What is a peeler? A crab about to shed its shell, usually recognizable by the light-colored line at the hind edge of the shell where the break will appear.

526. What is a sponge crab? A female crab carrying a spongy mass of eggs beneath her tail.

527. What is the purpose of the hard shell of a crab? The obvious purpose is, of course, protection of the internal tissues, but in addition to this the shell serves as an external skeleton to the inside of which the body muscles are attached. The hard shell material also makes possible the jointed structure of the legs, claws and other appendages which are responsible for the animal's locomotion, feeding and breathing, etc. Lastly, since the shell is periodically discarded and renewed, toxic waste substances produced by the body are accumulated as part of the limy salts composing the shell and so are removed.

528. Why do crabs shed their shells? Because, as with all other crustaceans and related creatures, the outer coat of the body is inelastic and does not stretch to accommodate growth. In order for the animal to grow, the shell must be cast off. Then the accumulated growth substances quickly absorb water and the body expands about 15 per cent, after which a new shell is laid down.

529. How much of the shell is shed at molting? All of it, down to the tips of the legs and the feelers, and including even the lining of the stomach.

530. How often does a crab shed its shell? It depends on how large and how old the crab is. When it is extremely small and young, and growth is rapid, shedding occurs every few weeks. It soon slows down, however, to once every few months, while in very large or old crabs it may be no more than once a year. If a crab became so large that it ceased to grow, it would have no need to shed. Growth, in fact, controls the time and frequency of shedding, although the control is chemical rather than mechanical.

531. How does a crab get out of its shell? A split occurs along the line at the back where the body turns under to become the tail. The split widens and the crab slowly backs out of its shelly skin, drawing its limbs and feelers and mouth parts carefully out of their cases, which are already detached.

532. Do crabs eat while in "soft-shell" condition? No. After shedding and before a new shell has hardened, even the mouth parts are soft and incapable of handling food.

533. What do crabs do while their new shell is hardening? Like any other tender creatures, they hide until strong or hard enough to face the world.

534. Are all crabs edible? So far as they have been tried. Whether a crab is eaten or not depends mainly on whether the amount of meat it contains justifies the effort of extracting it. Consequently, crabs have to be of fair size, either in body or claws, before they are thought to be edible. (See questions 522, 535.)

535. Which crabs are used as food? On the Atlantic Coast the principal commercial crab is the blue swimming crab (*Callinectes*) which is common from Cape Cod into the Gulf of Mexico as far as the Mississippi. The lady or calico crab (*Ovalipes*), which is another swimming crab with much the same range, is also used in the Southern States. Along the New England coast, however, the rock crab and the Jonah crab (species of *Cancer*), both of them red and slow-moving, are utilized to some extent. The stone crab (*Menippe*) is valued in the extreme south. (See question 554.) On the Pacific Coast, however, the red rock crabs (also species of *Cancer*) grow to much larger sizes, up to fifteen or sixteen inches across the carapace, and constitute the chief commercial source. In the case of Cancer crabs, most of the meat is found in the thick claws. The giant spider crabs of Alaskan and Japanese waters supplies much of the crabmeat which finds its way into cans. (See question 523.)

536. How do cancer crabs get their name? Not from any association with the disease cancer, but because this is the ancient Greek name for the edible crab of Europe, which in turn became one of the signs of the Zodiac, and also gave its name to the Tropic of Cancer marking the northern limit of the equatorial sun. (See question 505.)

537. What is the bubbly froth that the red cancer crabs produce at the sides of their bodies when stranded out of water? The froth consists of bubbles of air trapped in a slimy substance. Out of water, particularly when the crabs are disturbed or handled, the pumping action within the gill chambers goes on but air is pumped through instead of water. Hence the froth.

538. Are there any laws protecting the blue crab? A breeding sanctuary for crabs is set aside in Virginia waters between Hampton Roads and the Atlantic Ocean. Maryland prohibits the taking of crabs in November, at the time when the newly mated crabs are migrating southward to their breeding grounds.

539. How many eggs does a female blue crab carry? About two million.

540. Why are ghost crabs so called? Ghost crabs (*Ocypoda*), also known as sand crabs, are the color of the white, sandy beaches where they live, and they are to be seen one moment and not the next so that you wonder if you really saw one. They move very rapidly and stop very suddenly, so that they seem to disappear, blending with the color of the sand. They walk the beaches more at night and at dusk, when their flitting presence seems even more eerie. (See questions 514, 515, 516.)

541. Where do ghost crabs burrow? High up on the sandy beaches along the level of high tide, so that their burrows are cool and moist but never in danger of being flooded by water filtering through from the sea.

542. Why are fiddler crabs called fiddlers? Because of the extreme difference in size of the claws of the male, one claw resembling the fiddle or violin and the other the bow.

Male and female fiddler crabs (1″)

543. What is the large fiddler claw for? It is mainly for bluff or intimidation within a fiddler community. The male brandishes the claw, holding it aloft in a threatening or attention-getting manner either to attract or arrest a female or to fend off another male. Fighting is mainly between males. When real danger threatens, the crabs make for their burrows.

544. Do all fiddler crabs have a large claw? No. Females and immature crabs of both sexes have a pair of small claws.

545. Why do some fiddlers have the large claw on the left while the majority have it on the right? A male fiddler frequently loses its right claw in a fight with another male. When this happens a new claw regenerates, but while a new claw is growing on the right side, the original small left claw grows rapidly to become a large claw, and the new one on the right remains small. This is a quicker method of re-establishing the status quo (or its equivalent) than if a large claw had to be grown from scratch. (See questions 518, 519.)

546. What do fiddler crabs eat? They feed on whatever minute animals and plants the sand or mud contains. Instead of passing quantities of inert matter through the intestine, however, the fiddler crab selects such morsels as appeal to it.

547. Where do fiddlers make their burrows? They make long, slanting burrows about three feet in length, usually ending in a horizontal room. The entrances are made in the higher slopes of the beaches or mud banks, though below high-tide level. Before the tide rises high enough to flood the burrows, however, the crabs plug the entrances and manage to live comfortably in a fairly airtight and watertight compartment in a moist climate perfect for their needs. The burrows are lined with mud, which stops water from sifting in.

548. How do fiddler crabs dig their burrows? They dig holes in the sand or sandy mud by pressing the material between their legs and bodies and rolling it into balls, which they push out from the entrance of the holes. A newly made burrow can usually be recognized by the pile of small balls of mud and sand just outside the entrance.

549. How can fiddler crabs be caught in large numbers? Fiddlers are often used to feed terrapins in ponds. They may be caught by the bucketful on a beach by a number of persons running in a circle. As the hunters run, the crabs run too, and in the same circling direction, the circle slowly narrowing and the animal crowding more and more. The stamping and moving are effective; the shouting that the excited human usually adds to the action may relieve his own tension but probably adds none to the crabs'. When the crabs are bunched as closely as possible they are quickly shoveled into a waiting basket.

550. Do all fiddler crabs go in herds? Those of the East Coast—at least, those of the Middle Atlantic States—often go in herds of a thousand or more and can be driven like cattle herds, though a fiddler herd moves as a unit, more like a flock of sandpipers. Pacific Coast species, however, and the more southerly species on the Atlantic

Coast, do not have the herding instinct to any comparable degree and appear to scatter as individuals.

551. Where do fiddlers and ghost crabs spend the winter? Neither crab is much to be seen during the cold winter months, but arduous digging where their burrows are usually seen will show them deep in a sort of hibernating inactivity in the lower chambers just above reservoirs of salt water. (See questions 541, 547.)

552. How do fiddler and ghost crabs breed? The crabs carry their developing eggs as spongy masses beneath the tail as in crabs generally and—also following the general rule—the hatching young are set free in the water to start their life as free-swimming larvae. Even the land crabs living high in the hills of tropical lands go down to the sea to shed their larvae at certain times of the year. In Samoa, for instance, the annual migration of the land crabs to the sea to spawn takes place two or three days before the great spawning of the palolo worm (see question 722) and warns the natives to prepare for the feast to come.

553. How do fiddler and ghost crabs manage to live out of water? In spite of appearances, neither kind is a true air-breather. Each kind keeps a certain amount of seawater in its gill cavities and periodically dashes into the water to renew it. They can, however, survive for as long as six weeks without renewal.

554. What are stone crabs? Stone crabs (*Menippa*), are large, purplish-red crabs which burrow in sandy shoals or crawl beneath boulders in southern regions. They have very heavy claws and are good eating, and may be found from North Carolina south through the West Indies.

555. How do spider crabs get their name? Rather obviously, because their appearance suggests a spider. It is, however, not just a matter of long legs and small body, which are typical of some spider crabs, but of body shape. Spider crabs as a group have somewhat pear-shaped bodies which are narrow and pointed in front, so that the pairs of eyes and feelers are set close together. Legs may be quite short, as in the toad crab (*Hyas*) which lives on muddy or stony bottoms from Maine southwards; or they may be relatively long as in the

Spider crab (5–6")

common spider crab (*Libinia*) of the same territory. (See question 523.) To watch a toad crab feed is an education in itself, for as it sits with its wide bottom on the ground, its head up and its arms akimbo, passing food to its mouth with its utensil-like claws, it appears very efficient and much too human.

556. Where are kelp crabs usually found? Kelp crabs (*Pugettia*) are generally small olive-green spider crabs frequently found in rock pools along the Pacific Coast and very commonly clinging to strands of seaweed of the same color. The body and legs have very sharp spines which help the crab cling to the weed in the face of breaking waves. The claws are very strong for the same reason, and between

Kelp crab (1–2")

the claws and spines a collector needs to be careful how he handles the kelp crab. Large ones are best left alone. About one in twenty carries the parasitic barnacle (see question 664), the presence of which is evident only when the egg mass of the barnacle is extruded beneath the crab's tail. The mass has a much finer grain texture, is more compact, and is grayer in color than the egg mass of the crab itself.

Plumose anemones (Questions 788–801) and common starfish (Questions 166–168, 192)

Sun star (Question 192)

Sea urchin (Questions 202–218)

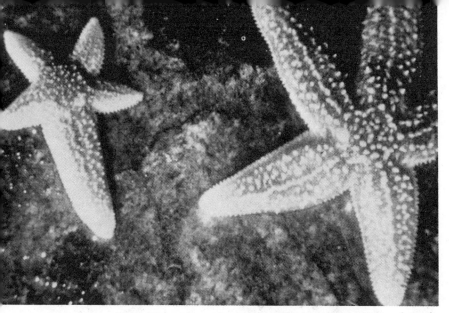

Common starfish showing sieve plate (Questions 160, 167, 192) and single ray regenerating four new rays (Questions 171–174)

Sea urchin shell or test (Question 215)

Brittle or serpent star (Questions 193–201) crawling over colonies of small sea anemones

Rock crab (Questions 505–537)

Green shore crabs mating (Questions 509–521)

Horseshoe crab (Questions 665–681)

Gooseneck barnacle showing stalk, protective shells and fan-like limbs (Questions 661–663)

Male fiddler crab (Questions 542–553)

Limpets in rock crevice (Questions 310–318)

Rock-purple dog whelks (Questions 262–269) feeding on acorn barnacles
(Question 659)

photo by J. B. Lewis

Sea snail showing crawling foot, mantle fold around base of shell, breathing siphon, tentacles and eye (Questions 246, 249, 289)

File shell or swimming clam (Question 406)

photo by J. B. Lewis

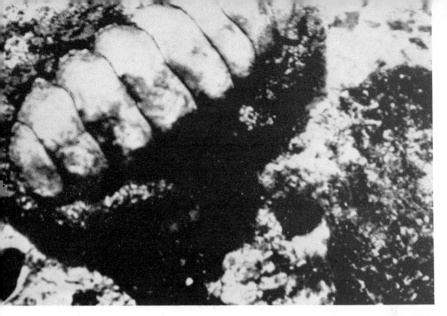

Chiton with eight shell plates and marginal fringe (Questions 447–456)

Spiny lobster (Questions 585–591)

Neck of long neck or soft shell clam showing intake and outgoing siphon openings (Question 395)

Inner surface of surf clam showing hinge and the marks of the two shell-closing muscles and mantle edge (Questions 369–397)

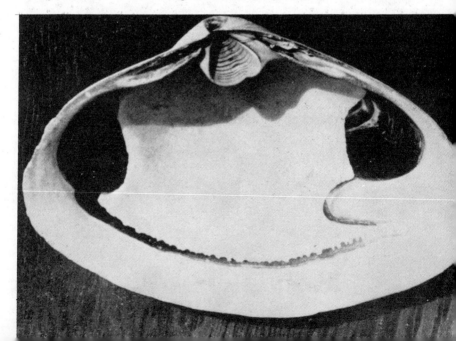

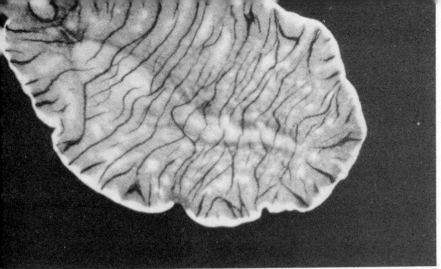

photo by J. B. Lewis

Flatworm (Questions 685–692)

Feather-duster worm showing crown of tentacles protruding from tube (Questions 754–760)

photo by J. B. Lewis

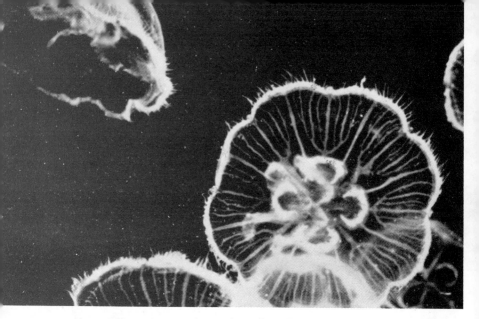

Moon jellies, showing horse-shoe-shaped reproductive glands, radial canals, and marginal tentacles and sense organs (Questions 810, 820–825, 832)

Hydroid colony of Tubularia, showing heads with tentacles and reproductive bodies (Questions 804–809)

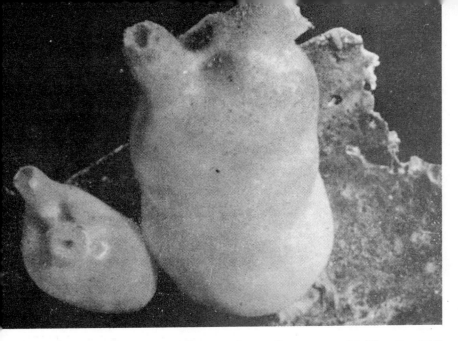

Sea peach or solitary sea squirt (Question 816)

Golden star colonial sea squirt (Question 918)

Herring gulls (Questions 1060–1067)

Loggerhead sponge (Questions 874–893)

Comb jellies showing eight rows of locomotory combs and pair of tentacles (Question 865)

Cormorants on cliff edge (Questions 1076–1082)

Head of young harbor seal (Questions 1032–1043)

557. What are pea crabs? Small crabs about the size and shape of peas, or a bit larger, with relatively small legs. There are many kinds, but all belong to one family. The majority are species of *Pinnixia,*

Pea crab (1/2″) in oyster

Pinnotheres, or *Scleroplax.* Most persons who have ordered oysters along the Atlantic Coast will have come across one of them, the oyster pea crab (*Pinnotheres ostreum*). As a group they specialize in living in the houses and eating at the tables of other creatures.

558. Where do pea crabs live? Certain species live within the tube of the parchment worm and in the tubes of various other worms. Other pea crabs live within the burrows of the mud shrimp, others again in the gill baskets of the larger sea squirts, while still others live only beneath the under surface of southern sand dollars. Each species of pea crab has its own preferred host. One species patronizes only a certain gaper clam, another mussels, another a large keyhole limpet, and so on. In all cases the host supplies safety, food and oxygen! (See questions 611, 228, 914.)

559. How do pea crabs feed? Little is actually known, since pea crabs obviously are difficult to observe. Many species, however, filter plankton or detritus out of the water the host brings into the tube. To some extent they feed at the expense of their host. "By putting glass windows in the shell of a mussel we were enabled to watch the activities of one of these pea crabs, *Pinnotheres concharum.* It obtains its food by eating some of the mucous strings by means of which the mussel carries food to its mouth."—(G. E. and N. MacGinitie.)

560. How do pea crabs breed? The pea crab that lives alone within the tube or shell of another creature is a solitary female. Yet she produces eggs and carries them in the usual way until they hatch and swim off as typical crab larvae. For a long time it was a mystery how mating was accomplished, for all crabs must mate in order that eggs may be fertilized. Moreover, most mature pea crabs have become too large to escape from their adopted homes even if they were so inclined. It now appears that the males are relatively small and are migratory animals with no permanent residence—in other words, typical small crabs wandering around as most crabs do. And in the course of their wandering they enter and mate with the ladies now imprisoned in their towers.

561. What is the difference between the common lobster and the spiny lobster? The common lobster (*Homarus*) has a pair of powerful, meat-filled claws. The spiny lobster (*Panulirus*) has no such

**American lobster
(Mature—9–20″)**

claws but only a pair of long legs in their place, not very different from the other walking legs. There are other differences but this is the most obvious.

562. Where does the common lobster live? Among the rock ledges off the shores of the Atlantic north of Cape Cod, particularly in rocky crevices below low water. It does not occur between the tides.

563. What color is the common lobster when alive? As a rule it is a dark greenish purple, although occasionally a light blue or green lobster is caught. They all turn bright lobster-red when boiled.

564. What do common lobsters feed on? They are scavengers and will eat anything from seaweed to dead fish. Live food is equally acceptable, and they have no aversion to eating mussels. They are too clumsy, however, to be efficient at catching active creatures.

565. How do lobsters breathe? As in crabs. (See question 509.)

566. What kind of blood do lobsters have? As in crabs. (See question 520.)

567. How large does the common lobster grow? The largest recorded weighed thirty-five pounds, but few grow to more than a fraction of this in these days of heavy fishing.

568. What is the legal size limit for taking lobsters? Ten and a half inches from the tip of the head to the tip of the tail, although the length of the carapace is what is usually measured. In any case it doesn't matter, since you will know the exact measurement if you are a lobster fisherman, and you shouldn't ask a fisherman or a generous friend embarrassing questions if you are not.

569. Can anyone catch lobsters? Not along the coast of Maine where the fishery is concentrated. Only permanent residents can obtain a license, and it requires three years to become a "Mainiac" for this purpose. A lobster fisherman, moreover, must catch his lobsters alone and unaided.

570. Why are the two claws of the common lobster so different from each other? The lighter claw is used for cutting up food and the heavier claw for crushing.

571. Can lost claws or legs be replaced? As in crabs. (See question 517.)

572. How are they replaced? As in crabs. (See question 518.)

573. Do the claws of the lobster grow at the same rate as the body? No. A young lobster has small claws relative to the size of the body, but as the body grows, the claws grow faster, so that in the thirty-five-pounder that made the record for size, the claws together weighed twice as much as the body.

574. Can lobsters swim? Only after a fashion. If in danger in the open, away from rock crevices, they flip their tails violently and shoot backward as a result.

575. How often does the common lobster shed its shell? Young lobsters molt from fourteen to seventeen times during the first year. A ten-inch lobster has molted about twenty-five times and is about five years old. Thereafter it molts but once a year or perhaps only once in two years.

576. What is the purpose of the hard shell? As in crabs. (See question 527.)

577. Why do lobsters shed their shells? As in crabs. (See question 528.)

578. How does a lobster get out of its shell? As in crabs. (See question 531.)

579. How much of the shell is shed at molting? As in crabs. (See question 529.)

580. What enemies do lobsters have? Apart from man, their chief enemy is the cod, which destroys large numbers of the smaller lobsters.

581. What is meant by lobster in berry? One that is carrying her eggs.

582. How many eggs does the common lobster carry? In every alternate year an eight-inch lobster carries about five thousand eggs, a twelve-inch lobster about twenty thousand, and a fourteen-inch mature matron as many as forty thousand. It pays to let them live.

583. For how long does the common lobster carry its eggs? The developing eggs are carried attached to the appendages beneath the tail for ten to eleven months.

584. Do lobsters stay in the same place all year round? No. Young, newly hatched, lobsters swim near the surface of the sea for nearly two months, after which they sink to the bottom and for the first time look like small lobsters. They remain in shallow water along the shore through the summer and, at least when fairly well grown, migrate each fall into deep water offshore and return inshore each spring. (See question 511.)

585. Where do spiny lobsters live? Spiny lobsters, also known as sea crawfish, are a southern form common on coral reefs and under rocks in lagoons throughout the West Indies and Florida, and also on the Pacific Coast from Point Conception southward.

586. What are the colors of the live spiny lobsters? They are varied, with patterns ranging through dark brown, red, yellow and some bluish regions.

587. Why is the spiny lobster so called? Because the shell is provided with numerous needle-sharp spines, which are an important

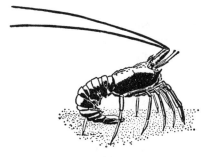

Spiny lobster or crawfish (12–20″)

means of defence. Apart from the spines and the armored shell itself the spiny lobster is virtually defenceless.

588. What is the noise a spiny lobster makes when handled? There is a special noise-making apparatus at the base of each of the two long feelers, which makes a grating noise. They only make the noise when molested, and it is possible that the noise may scare off an attacking fish. The lobster itself has no known sense of hearing and it may or may not be aware of the noise.

589. What do spiny lobsters feed on? Like the common lobster they are mainly scavengers, although they have a taste for small snails and at night search the eelgrass and turtle grass, crushing the snails with their mandibles below the mouth.

590. For how long does the spiny lobster carry its eggs? For about two and a half months.

591. How many eggs does the spiny lobster carry? About half a million. The eggs are smaller than those of the common lobster. (See questions 539, 582.)

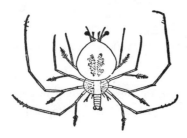

Drifting larva of spiny lobster (1/2″)

592. What kinds of shrimp are there? There are several very different kinds, including the true shrimps and prawns (*Crago, Pandalus, Penaeus*) which are fished commercially (see questions 595,

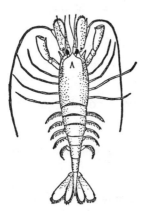

Common sand shrimp (2–3″)

Commercial shrimp or prawn (4–6″), carrying eggs

600), together with the pistol shrimps (*Crangon, Synalpheus*) and the broken-back or tide-pool shrimps (*Hippolyte, Spirontocaris*) (see question 599), and certain others such as the ghost shrimps, which are a very different sort. (See questions 608–611.)

593. Is there any difference between a shrimp and a prawn?
Prawn is a name given to the larger kinds of swimming shrimps.
Shrimp is the name given to all small kinds whether they are swimming types or not. The common prawn (*Palaemonetes*) is abundant
among eelgrass and in ditches and salt marshes from New Hampshire
south.

594. How do the prawns or swimming shrimps swim? Apart from
backward movement produced by flips of the tail, they swim by means
of the processes attached to the underside of the tail. There are five
pairs of these appendages, each pair locked together as though holding hands, and when swimming the five pairs beat in rhythmic succession, causing the animal to move forward. (See question 573.)

595. What shrimps are fished commercially? On the Pacific Coast
the San Francisco shrimp and the coon-striped shrimp of Puget Sound
are fished to some extent, but the bulk of those that reach the market
come from the Atlantic Coast and the Gulf of Mexico. The East
Coast form is the swimming shrimp or prawn, and the shrimp fishing
is carried on from North Carolina to Panama.

596. How long do shrimps live? Whether they grow to be one inch
or six inches, in most cases their life span appears to be little more
than one year. The Puget Sound coon-striped shrimp, however, lives
between two and three years.

597. How do shrimps breed? In the case of those most fully investigated, the shrimps go offshore into deep water to hatch the eggs,
which are attached beneath the tail as in lobsters. By the time the
young look like shrimps, they are found in bays and creeks with shallow brackish water and muddy bottoms. Then as they grow they move
slowly back to deeper water of high salinity. (See questions 511, 551.)

598. Can shrimps change sex? Some do and some do not. The large
commercial East Coast shrimps are, as individuals, male or female
throughout their life, the males reaching an average length of six
inches and the females nearly an inch longer before they die off. On
the other hand, certain shrimps of the Atlantic Coast change sex as

they grow, being first male and later female. The coon-striped of the Pacific Coast is similar, the males changing to females when about two years old. (See questions 416, 443.)

599. What are the tide-pool shrimps? Tide-pool shrimps are mainly the beautiful little broken-back shrimps, although they are likely to be found among eelgrass as in the lower tide pools. They are called broken-backs because of their disconcerting manner of suddenly darting backwards by suddenly flexing the tail. Some are greenish or

Tidepool or broken-back shrimp (1–1 1/2″)

brownish, spotted or banded with red, and in general so colored as to blend with their surroundings. They are translucent, which makes them even harder to see, although when captured and closely examined you can see the heart beating and the workings of the stomach, etc. At night their pigment cells contract and the delicate creatures become virtually transparent. (See question 592.)

600. What do sand shrimps feed on? The sand shrimps (*Crago*) are among the most proficient scavengers along the shore, feeding especially on the debris churned up by the waves at the water's edge. They are hard to see, since their color is that of the sand, but any barefoot wader is familiar with the feel of their small bodies shooting across his feet as they try to get out of the way. (See question 640.)

601. What are pistol shrimps? The pistol shrimps, or snapping shrimps, are small but remarkable shrimps that have a very large

Pistol shrimp (1″)

pistol claw equipped with a snapping device which is most efficient as a defensive or offensive weapon. (See question 543.) If the claw is lost, the small claw grows large instead. (See question 545.)

602. Where can pistol shrimps be found? On the Atlantic Coast they may be found about the oyster beds and elsewhere, but one species is a regular inhabitant of sponges from North Carolina southward. On the West Coast they may be taken from under rocks at the lower intertidal level and from among sponges and sea mats in the tangle of kelp holdfasts. (See question 103.) As a rule they are heard more readily than they are seen. The snapping of a claw within your hand causes a stinging sensation.

603. Why do pistol shrimps snap their big claws? Not for the sake of making a noise which they probably cannot hear, but for the stunning effect produced by the sudden action. "When seeking food, a pistol shrimp lies just at the entrance of a burrow with its long antennae extended to detect any passing movement. When a small fish passes by, a shrimp slowly creeps out, and, before the fish is aware of the presence of the shrimp, it is paralyzed by a 'shot' from the pistol hand. The shrimp then quickly reaches out, grasps the stunned prey and drags it inside the burrow, then proceeds to make a meal. . . . We have also watched pistol shrimps hold a worm or other intruder in the small hand while the prey was snapped and paralyzed with the large hand."—(G. E. and N. MacGinitie.)

604. How loud a noise does the pistol shrimp make? If the shrimp is in a glass aquarium or dish, the noise will be heard all over a large room.

605. How do pistol shrimps make the noise? By snapping the thumb of the big claw against the palm with great suddenness, like a man snapping his finger. The sudden force is made possible by a trigger at the joint.

606. How strong is the snap of a pistol shrimp? When snapped against the wall of a battery jar used as an aquarium, it has broken the glass when the latter has been previously scratched by having been dragged on a cement table.

607. How do pistol shrimps shed the shell of so large a claw? The "hand" is so large in comparison with the size of the body, and the leg which joins it to the body is so narrow, it is quite impossible to pull it out. Consequently, when the shrimp molts, the shell of the large hand and leg shakes off in flakes instead of the limb being withdrawn in the usual way. (See questions 578, 527.)

608. What are ghost shrimps and where are they found? They are mud-dwelling shrimps (species of *Callianassa* and *Upogebia*) two or three inches long with one very large and one very small claw, that

Ghost shrimp (2″) in burrow

live in burrows in bays and marine sloughs from Alaska to Lower California and from Cape Cod to South Carolina. They live on bottoms that are not too muddy or too watery or too sandy—in other words, the bottom must be just right! In the right place, however, they can be very abundant even between the tide marks.

609. Why are they called ghost shrimps? Because of their whitish-yellow color. Since they spend most of their life below the surface of the mud, no protective coloration is needed.

610. How do ghost shrimps live? A life of hard labor! They have to dig to build their homes, and have to dig to eat. Food is obtained from the organic material in the mud, which has to be sorted and sifted to get a meal. It is dirty work, and only the second and third pairs of legs (the first being the claws) are used in digging, while the fourth and fifth pairs have become long-handled brushes for cleaning mud off the rest of the animal, and its eggs if it should be carrying them. The burrows have two openings at the surface, and a current of water is maintained through it. (See questions 547, 707–712, 772.)

611. Who else lives in a ghost-shrimp burrow? In most cases a ghost shrimp (all species) is host to a scale worm and a pea crab. (See question 559.) But the most remarkable species is the Californian ghost shrimp that harbors a pair of blind fish, the blind goby. (See question 998.)

612. What is the large claw of the ghost shrimp used for? It does serve as a weapon for offense or defense, but only the male possesses it and the female with her smaller claw seems to get on just as well. Possibly the male ghost shrimps have inherited their large claws from a time when they spent their lives above ground, and now simply put up with the cumbersome antique. (See questions 543, 603.)

613. What happens if a ghost shrimp is removed from its burrow? If it cannot make or find a new one at once it dies within a few hours. Apparently life is intolerable unless it can feel the burrow walls around it, even when all other conditions are satisfactory.

614. Where are mantis shrimps found? The mantis shrimps, so called because they hold their wicked fore-limbs in the "ready" position suggestive of the praying mantis insect, are to be found on both coasts—on the Atlantic Coast from Cape Cod into the West Indies, in shallow burrows in mud or between coral rocks between and below the tides; and on the Pacific Coast south of Point Conception, deep in

Mantis shrimp (3–6")

burrows below the lowest intertidal rocks. They are almost six inches long. The Atlantic form (*Squilla*) is often seen as a fast streak moving from one shelter to another in very shallow water, extremely difficult to catch, while the Pacific form (*Pseudosquilla*) requires even greater effort to collect. In either case, if successful you find a fascinating creature to look at, but one that can—and most likely will—give you a very nasty nip if not a severe wound.

615. What are the colors of the mantis shrimp? The Atlantic form is greenish, edged with bright yellow, but the California species is said to be one of the most beautiful animals in the sea. Its body is a combination of azure blue and orange, with a deeper-blue tail edged with light blue and gold, and the whole invested with a delicate translucent quality.

616. What are the weapons of the mantis shrimp? Mantis shrimps have what is called a "jackknife claw," because the outer blade fits into a groove in the inner position just like a jackknife blade. They make slashing movements so fast that the motion can scarcely be seen, and can cut a shrimp in two as though it had been guillotined.

617. Where are hermit crabs found? Hermit crabs (*Pagurus*) are generally much alike except in size and color. On the Atlantic Coast a small kind is one of the most abundant water's-edge creatures from Maine to Florida. It is found everywhere on sandy and muddy bottoms, in rock pools and behind sand bars, etc. A much larger reddish-brown kind also extends from Maine to Florida but in a little deeper water. (See question 32.) A purple-clawed kind lives among the mangrove keys.

618. Why do hermit crabs need a shell? This is somewhat like asking which came first, the chicken or the egg. Hermit crabs as they now exist have soft, unprotected rear ends curved in a spiral which fits a snail shell. Without the snail shell they are quickly torn apart by other crabs. Probably their shrimp-like ancestors acquired the habit of sheltering in empty snail shells and in the course of innumerable generations their descendants became so fitted to the shell they can no longer do without. (See questions 527, 621.)

619. Do hermit crabs use only empty shells? Yes. They are not capable of extracting the original mollusk and must wait for shells to be empty.

620. Do they clean out shells before using them? No, but they make sure they are all right before occupying them. A hermit in need of a new shell reaches down inside it to find out whether it is clear of

Hermit crab (1–3″) exploring empty moon snail shell, and installed

other crabs, worms or debris. Only if everything seems satisfactory does it make the change, and then as fast as it possibly can.

621. Are hermit crabs always found in snail or whelk shells? Almost all hermit crabs occupy the empty shells of snails and whelks. One species from deep water in the Indian Ocean, however, inhabits pieces of bamboo, possibly because beggars can't be choosers. The other exception is the coconut crab, a veritable giant which has taken to the land and has dispensed with mollusk shells, having succeeded in rehardening its own skin. (See questions 527, 618.)

622. Do hermit crabs always use the same kind of shell? They are not fussy as to whose shell it was, but the size of the shell is all-important. Of the two kinds of hermits common along the Atlantic Coast, the larger one inhabits the shells of the larger whelks and moon snails, while the smaller kind is generally found in the shells of periwinkles and mud snails. On the Pacific Coast the turban shells are commonly used. Availability and size are the important matters.

623. Why does a hermit crab have to change its shell? For the obvious reason that it keeps on growing, while the dead mollusk shell it occupies does not. (See question 527.)

624. What happens when it grows too big for its shell? It must find a new shell that is both unoccupied and somewhat larger. If after moving into a new one it is found to be unsuitable, the animal quickly changes back and proceeds to look for another. (See question 620.)

625. Do they ever choose a shell too big to pull around? As a rule they drag around shells which are proportionate to their size and power, but occasionally a hermit crab gets into one which is so loaded down with barnacles and slipper limpets that progress is almost impossible. Probably if no progress at all could be made, the animal would eventually get out and look for another, for even a hermit crab is known to be able to learn from experience.

626. How do hermit crabs protect themselves? By withdrawing as far as possible within their shells. Small kinds may withdraw almost out of sight, while larger kinds have relatively larger claws which block the entrance of the shell like an armored shield.

627. What do hermit crabs feed on? They are primarily scavengers but will eat any flesh they can get hold of, including any of their own kind that are too slow in getting into a new snail shell. (See question 620.)

628. Why are hermit crabs so hard to pull out of their shells? In the first place the body of a hermit crab is a tapering spiral which fits the spiral curve of the inside of the snail shell. Equally, if not more important, the last pair of limbs on the abdomen serve as a clamp which secures the animal to the pillar of the shell. This was observed by keeping hermit crabs in glass models of snail shells.

629. How can hermit crabs mate? Males are generally larger than females and as a rule there is considerable competitive fighting for possession. During the mating urge a male often drags the shell of a female around waiting for her to shed her skin, after which he deposits sperm within the mollusk shell onto her abdomen where it will fertilize the eggs as they are laid. (See question 521.)

630. How do hermit crabs take care of their eggs? The eggs are carried by the appendages of the abdomen or "tail," as in the case of other crabs etc., and number many thousands. Owing to the curve of the body they are carried only on the left side. When there is no danger, the female often comes far enough out of the shell to wave her egg masses too and fro in order to clean and aerate them. When they are ready to hatch, she partly emerges again and brushes them off with a brush on the last leg of her left side. (See question 521.)

631. How big are they when they first acquire a shell? When they first settle on the sea floor as miniature hermits, they are but a small fraction of an inch long. Since small, empty mollusk shells are plentiful they have little trouble finding one that suits. The difficulty increases as the hermits grow larger.

632. Are hermit crabs ever used as food? Italians regard them as delicacies when cooked in oil. They are served in the shell and taken out with a pin as they are eaten. Everyone to his taste! (See questions 535, 595.)

633. What grows on hermit-crab shells? A snail or whelk shell occupied by a hermit crab is a very suitable place for many other organisms, for a variety of reasons. Barnacles are frequently found on the outer surface, although they may well have been there when the shell was occupied by its molluscan maker, while slipper limpets tend to settle on the smooth inner surface of the large kinds. (See questions 649, 326.) The more spectacular guests, however, are specialists in this sort of visiting. The fuzzy, pink-and-orange hydroid (*Hydractinia*) forms a velvet coat on a considerable percentage of the shells of small hermits of the Atlantic Coast, while the Adamsia anemone often envelopes the shell of the larger kind in southern waters. (See questions 809, 803.)

634. What are mole crabs? Mole crabs, (*Emerita* or *Hippa*), also known as "sand bugs" and "sand crabs," have more the shape of a lobster with its tail well tucked under than that of a typical crab.

Mole crabs burrowing (1")

They are, in fact, more closely related to the hermit crabs and ghost shrimp and none of them are true crabs. The name comes from their mole-like shape and their ability to dig into the sand in a most effective manner. The body is usually about an inch and a half long.

635. Where can mole crabs be found? Along many beaches of both Atlantic and Pacific Coasts, usually at the edge of the tide where the waves are breaking. They burrow only in fine clean sand, digging in backward so that only the V-shaped pair of antennae are left exposed. If they are left stranded by an ebbing tide, they dig three to six inches below the sand and await the return of the tide.

636. Do mole crabs stay in one place? No. They burrow but have no burrows. As a wave travels shoreward the crabs leave their burrows and allow themselves to be carried along by the water until the wave slackens, then they dig into the sand again. This may happen several times during the course of an incoming tide, and is repeated in reverse when the tide is ebbing. As a rule, whole communities will shift at the same time.

637. How do mole crabs feed? They feed by means of their projecting antennae, which collect minute food particles from the film of water receding after each wave. The food is scraped off the antennae between one wave and the next.

638. Do mole crabs have any enemies? Principally shore birds, but also human bait collectors who take them in large numbers when in their soft-shelled condition.

639. What is the gribble? A very small crustacean (*Limnoria*), which burrows into submerged wood, particularly into wharf piling at the low water level, feeding on the wood and raising its young within

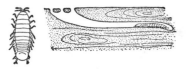

Gribble (1/4"), showing tunnel
in wood with openings

the burrows. Eventually the piling is completely eaten through at the level of the mud or sand, and the supported structure collapses. (See questions 405, 432–443.)

640. What causes the stinging sensation between your toes if you stand too long in one place in the water along the sand beaches of Southern California? A small crustacean (*Cirolana*) that is related to the sow bug or wood louse which lives under the bark of rotting wood on land. The bite is uncomfortable, no more, and the animal is merely mistaking your toes for dead mole crabs. Its close relative the gribble, however, is a most destructive creature in connection with pilings and other submerged wooden structures. (See question 639.)

641. What are sand fleas? They are not fleas at all, which are insects, but are marine crustaceans better known as "sand hoppers."

They feed for the most part on decaying seaweed and are perfectly harmless.

Beach hopper or sand flea (1/2")

642. Where are sand hoppers found? They are found most abundantly under piles of rotting seaweed tossed up along the beaches at the level of high tide, from which they get their food. Some kinds live more among the grass and weeds of salt marshes, while close relatives such as the side-swimmer (*Gammarus*) is most abundant under stones near low water. (See questions 135–136.)

643. How do sand hoppers hop? Sand hoppers (*Orchestia*), also called "sand fleas," jump considerable distances by using their tails and last three pairs of legs as a spring. (See question 641.)

644. What are skeleton shrimps? Skeleton shrimps (*Caprella*) are not shrimps but are related to the common sand hoppers. They are slender creatures up to an inch or so long which move about among

Skeleton shrimp (1") in fishing attitude

hydroids and seaweed much in the manner of a measuring worm. When not looping along, they hold on by their hind feet and stretch out in the water with a pair of pincer-like claws held ready to grasp any unwary morsel that may swim too close. Fortunately they are small, for few living creatures have such a fantastic appearance. If they were big they would scare us to death.

645. What are barnacles? All appearances to the contrary, barnacles are crustaceans with as much right to the name as has a crab or a lobster. To get the picture you have to imagine a crab, or some such creature, lying on its back with legs in the air and surrounded by a

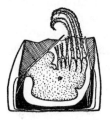

Acorn barnacle (1/2″) showing animal inside shell

fortress of limy plates. A barnacle is in fact a typical crustacean which has become permanently fastened by its head to a rock or some other solid surface and has in consequence lost its sensory equipment of eyes and feelers.

646. Where do acorn barnacles live? For the most part the acorn or rock barnacles (*Balanus*) occurs in fairly definite zones with relation to the tide, some forms thriving at the highest levels, some more successful further down, and some only below low tide. Those in deeper

Acorn barnacles (1/4–1/2″), showing feeding individual and empty and closed shells

water are generally more or less solitary and of relatively large size. Acorn barnacles also grow freely attached to wharf piles and ship bottoms, not to mention the shells of mollusks and crabs etc. (See questions 633, 670.)

647. How abundant are acorn barnacles? Any casual glance at a barnacle-covered rock will give you the impression of great abundance, but the actual figures are startling. On a thousand-yard stretch of one rocky shore roughly one thousand million were estimated to be present, which would produce around one million million larvae in the course of a year.

648. What kind of barnacles are there? The two principal kinds are the acorn barnacles, which are squat forms somewhat the shape of acorns, and the gooseneck barnacles (*Lepas*) which have long flexible stalks, the head and neck suggesting the head and neck of a goose. A third kind, which does not look at all like barnacles, is a group of most peculiar parasites. (See question 664.)

649. How can acorn barnacles best be watched? Either by lying down on the rocks with your face practically in a tide pool or by putting a stone with some barnacles attached to it in a dish of cool, clean seawater. Then, with or without a hand lens, you can see the feathery feet reaching out of the small white house as it fans for invisible food in the water. It is a sight worth seeing.

650. How do barnacles feed? All kinds feed in essentially the same way. The six pairs of appendages comparable to the locomotory limbs of the other crustaceans here act together as a hand-like feeding organ. When feeding, the protective shelly valves separate and the "hand" flashes out and draws its fringed fingers through the water in order to trap any food particles, living or dead, that it may contain. The process is repeated regularly every few seconds. (See question 637.)

651. How do acorn barnacles reproduce? Like all barnacles they are hermaphroditic—that is, have both sexes in the same individual, although one barnacle fertilizes another and is fertilized in turn. Each has a long, slender extensible sperm tube or penis which can be protruded through the valves of the shell, even to reach other barnacles several inches way. The eggs develop within the protective chamber of the parent and hatch out as microscopic larvae similar to those of the majority of crustaceans. The larvae swim and feed in the open sea for

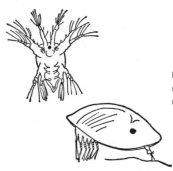

Larval stages of acorn barnacle (1/20–1/10")
showing freeswimming stage and stage exploring,
rock surface for suitable spot to settle on

several weeks before they sink and search out a suitable place to attach to. (See questions 629, 630.)

652. How do the larvae of acorn barnacles find a suitable spot to settle on? By the time the larvae are ready to give up their freedom
for a sedentary life they already possess six pairs of legs, a pair of
antennae and a hinged, bivalve shell. This stage does not feed and is
solely concerned with finding the right place for future living. It sinks
to the bottom as the tide carries it in, and explores the ground
for about an hour with its antennae before it makes a final choice.
Rough surface, shade and water velocity are the chief items which
influence the final settlement.

653. How do acorn barnacles fasten themselves to rocks etc.? The
larva fastens itself to the surface by means of cement produced by a
special gland associated with its antennas. Thereafter more cement is
added as needed and at the same time a limy base is formed below and
around the soft tissues of the animal.

654. How do acorn barnacles grow? In the same general manner
as other crustaceans, that is, the body must cast off its old skin before
it can expand. Then a new coat is formed. All this however takes
place within the safety of the limy shell. At times the water in bays
where barnacles are common appears full of their discarded molts,
each one looking like the cast-off glove of some small, ghostly mani-
kin. The outer protective shell of lime, which is usually made up of
six plates joined together, enlarges by addition of new material to the
edges of each plate, although in a somewhat complicated manner.

655. How large do acorn barnacles grow? Most intertidal kinds rarely exceed one quarter or one third of an inch across, although they may be an inch high when crowded together. Probably the largest in the world is a lower-level form which ranges from southern Alaska to Lower California. It is commonly three inches high and even wider at the base, while under some conditions it is reported to be six inches high and three to four inches wide, weighing a good half pound. The soft mantle through which the feeding legs protrude in this form is highly colored with bright reds and purples. It is usually seen attached to wharf pilings, though it also attaches to rocks below the low tide level.

656. Why are some acorn barnacles broad and squat and others tall and narrow? Barnacles attached to rock surfaces where there is plenty of room tend to spread out, but when they are crowded together from the start and press against one another, the only way they can grow is upward, so that they come to look like extracted teeth.

657. How do acorn barnacles withstand exposure when the tide is out? They not only have the impenetrable wall of stony plates enclosing most of the body but also a pair of tiny valves which can be closed so tightly that the water inside is retained until the tide returns. (See question 266.)

658. How long do acorn barnacles live? Those high up on the shore, which are slower getting into their growth stride, live about five years; those farther down, which grow fast from the first, usually die at the end of their third year.

659. What enemies do acorn barnacles have? Various kinds of small snails and whelks feed extensively upon barnacles, although not so much that their numbers are significantly affected; they easily account for the number of empty shells one finds attached to rocks. The dog whelks such as the common rock purple and dye shells live alternately upon barnacles and mussels. (See question 265.)

660. If the two valves of an acorn barnacle are closed tight, is the animal alive? As a rule. When the barnacle dies, the valves open and the tissues decay; or when a snail eats a barnacle it forces the valves open. Empty barnacle shells are wide open at the top without even a sign of the valves, and look like miniature volcano craters.

661. Are barnacles ever eaten? Most barnacles are far too small to justify the effort of extraction, but giant West Coast barnacles are still eaten by the Indians of the Northwest and also by other people. The flesh is said to taste very much like that of lobsters and crabs. In the case of the gooseneck barnacles only the neck or fleshy stalk is eaten; it is favored by the Spanish and Italians.

662. Where are gooseneck barnacles found? A West Coast form (*Mitella*) is common on the rocks in the mussel zone from Alaska to Lower California, but gooseneck barnacles are more typically voyagers. The common goosenecks (*Lepas*) are usually found attached to pieces of floating wood, ships bottoms, seaweed or glass bottles, often far out to sea but frequently drifting ashore with wind and tide. One of them even has a self-manufactured float which enables it to dispense with artificial aids to a drifting existence, while another kind lives attached to a large acorn barnacle which itself lives attached only to the skin of whales.

663. What is the stalk of the gooseneck barnacles? The stalk is an extension of the front of the head and corresponds in part to the original antennae of the larvae. The cement organ used for attachment is situated at its end (see question 653). In medieval times the stalk increased the mystery as to what barnacles really were and it was thought that they grew on trees and gave rise to "barnacle-geese," which some people considered to grow into real geese.

664. What are parasitic barnacles? They are typically barnacles when very young up to the time they settle and become attached. The attaching stage, however, lives only if it fastens itself to the back of a crab or certain other crustaceans. Then when the crab molts, the barnacle animal works its way inside the crab's tissue, loses all resemblance to a barnacle and looks and grows like an invading malignant tumor. When the crab begins to weaken, the parasite (*Sacculina*) produces a large egg mass of its own which extends outside the crab between the body and the turned-under tail. The parasite eggs develop into barnacle larvae and those that settle on other crabs carry on the infestation. The parasite is common on the green crab of European shores but for some reason is not found on the same crab of the Atlantic Coast of North America. However, about one in twenty of the small kelp crabs (see question 556) of the rock pools along the Pacific Coast carry the parasite. The egg mass can be distinguished from that of the crab by its firm texture and grayish color.

ARTHROPODS—HORSEHOE CRABS

665. Are horseshoe crabs really crabs? No. Horseshoe crabs (*Limulus*), also called "king crabs," are not even crustaceans. Their closest relatives now alive are the scorpions and spiders of the land, although horseshoe crabs have never been land forms themselves. According to fossil evidence they have persisted virtually unchanged for more than two hundred million years and are often called "living fossils."

666. Where do horseshoe crabs live? In sheltered shallow waters below low tide, where they crawl on mud and sand. Large individuals are readily seen when wading, while small ones may be trailed along the extensive sand flats left at low tide along the coasts of the Caro-

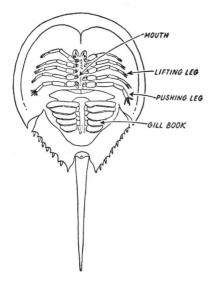

Under surface of horseshoe crab (2')

linas. The range of the horseshoe crab is from Maine to Mexico on the Atlantic Coast. They are not found on the American Pacific Coast but are common along the coasts of southeast Asia.

667. Are the horseshoe-crab shells you find along the shore dead horseshoe crabs or only the cast shells? They are the cast shells. (See question 679.)

668. What kinds of movements can the body make? The horseshoe-crab body is made in three parts. The great shield is attached to the abdomen by the strong hinge which works only up and down. The long spine-like tail is joined to the abdomen in a ball-and-socket joint, so that it has universal movement.

669. What is the tail spine used for? The spine of a horseshoe crab is not a weapon either of offense or defense but is used as a lever for righting the animal whenever it gets turned over on its back by the waves. The tip is then thrust into the sand and the creature arches itself over. Unless the point can get a grip the horseshoe crab is helpless.

670. What is the purpose of the great shield of the horseshoe crab? It is protection against other animals and it also serves to keep the creature right-side-up in the rough waters of the tidal flats. Most of the time the horseshoe crab progresses with the flange of the shield imbedded in the sand or mud like the biting edge of a plow, so that the force of waves merely presses the animal down into the ground and cannot get under the shield to lift it over. The topside of the shield is often covered with acorn barnacles (see question 646) and the underside often has slipper limpets attached to it. (See question 326.)

671. How do horseshoe crabs move? The progression is peculiar. Four pairs of walking legs are alike, each leg ending like a claw—which, however, serves more like the cloven hoof of a deer for getting a grip. These four pairs of legs lift the heavy body off the ground, at least to some extent. Then the last or fifth pair comes into play. This pair is larger than the rest and gives the body a violent shove forward so that the animal gives a lurch. This last pair is made like a pair of ski poles, for their ends are equipped with a central projection which penetrates the sand, and a circle of lateral projections to stop it from going in too far.

672. How do horseshoe crabs feed? Each of the ten legs has a large base equipped with short, close-set spines. As the animal progresses, the five pairs of bases push together and grind up any food that lies between, which then passes up into the mouth. The food is obtained by plowing slowly through the sand or mud with the flange of the shield and stirring up the material. Any small creatures which are churned up are seized by a pair of small claws in front of the legs and passed back to the leg grinders. The whole apparatus, however, is like a piece of automatic machinery, for the animal cannot eat unless it is moving, the two actions going together.

673. What do horseshoe crabs feed on? Small bivalves and worms which are found living just beneath the surface of the sand or mud.

674. Where are the eyes of the horseshoe crab? There is a large one on each side of the shield which is fairly obvious, and another though smaller pair situated up front on either side of a small medium projection of the shell.

675. Where is the mouth of the horseshoe crab? On the undersurface in the middle between the bases of the five pairs of legs.

676. How do horseshoe crabs breathe? They have unique structures attached to the underside of the abdomen, called gill books, which may be compared to the swimming appendages beneath the tail of a shrimp. There are six pairs, the first pair acting as a cover for the rest and called the operculum. The other five pairs are the respiratory organs, and each of the ten books constituting the five pairs consists of about eighty pages or broad sheets through which blood flows. The books are shaken open in the water in a regular rhythmical manner so that oxygen is exchanged. (See question 509.)

677. What kind of blood do horseshoe crabs have? Blue blood, like that of crustaceans and most mollusks, with the same copper base.

The properties of blue blood have been more closely investigated in the case of the horseshoe crab than in any other creature. (See question 520.)

678. How large do horseshoe crabs grow? When fully grown, the females are about twenty inches long. Males do not grow quite so large.

679. How does a horseshoe crab get out of its shell when molting? The shield splits along the whole of the front edge and down its sides. The animal then squeezes forward out of its shell, in contrast to true crabs and lobsters which pull out backward. When the horseshoe crab has completely shed its shell, the old shell practically closes up and looks like an entire animal. (See question 528.)

680. How do horseshoe crabs breed? In early summer the females move out of the water and up the beach, the males following close upon their tails. Deep depressions are made in the sand and the fertilized eggs, each about the size of a frog's egg, are laid and covered over. Several weeks later the eggs hatch out and small creatures looking like miniatures of their parents, minus only the spiked tail, are washed down by the tide to start life on their own.

681. Have horseshoe crabs any use? They are not eaten but have been used extensively in the past as fertilizer on farms along the Atlantic Coast, with a consequent diminution of their numbers.

682. What are sea spiders? Wherever you find thick masses of hydroids or groups of sea anemones you are likely to find the so-called sea spiders (*Pycnogonum*) whose larvae feed upon their juices. The animals look remarkably like spiders and may be distantly related to them—grotesque creatures with eight jointed legs and little body, with a span up to an inch or more. The adults are generally found crawling about under stones, in rock pools, or among seaweed below the tide.

683. What are the little blue-black bugs that cluster in large numbers on the surface of the water of high-level tide pools? They are wingless insects (*Anurida*) which are never found above high tide nor below low tide, and in so far as they can be called marine they are among the only two or three marine insects in the world. They are abundantly present and may be found not only on the surface film of the water but also crawling over seaweed and under rocks. They can breathe under water because they have a thick coating of hair which contains enough air for several days' use.

WORMS

684. What is a worm? The term can be used very broadly to include a number of separate groups of animals, such as the flatworms, ribbon worms, segmented worms, and some others, or in a restricted sense to indicate only the so-called true or segmented worms. Generally speaking, however, the name is given to any creature which is long and slender and without obvious appendages for locomotion, so that some mollusks (e.g., the shipworm) and even some reptiles (e.g., the slowworm) are called worms because of this superficial appearance.

685. What are flatworms? As the name implies, they are relatively long and narrow but are also flat, so that the shape cf the flatworm is very much like that of a leaf. It has a head end and an upper and

Flatworm (1–2″)

lower surface. It may be several inches long or no more than half an inch, and may be vivid in color, with striking pattern, or drab and inconspicuous.

686. How do flatworms feed? The animals are carnivorous, and in all the larger marine kinds the mouth is well back on the undersurface near the middle of the body. It is the only opening into the gut and, as in the case of the sea anemone and its relatives, any undigested material must leave by the mouth. Encircling the mouth there is a frill-like fold capable of great extension, and when the animal is feeding this is extended and drawn around the food material as a sort of external stomach into which the powerful digestive juices are poured. When sufficiently broken up by chemical action, the food is sucked into the interior. (See question 179.)

687. What do flatworms feed on? All are flesh-eaters and most are catholic in their choice of food, feeding on small worms, snail eggs or anything else that they come across. They have been known to eat segmented worms four times their own length. Flatworms, however, are not scavengers but take their food alive; different species have at least some preferences of their own. Many concentrate on the microscopic organisms that they crawl over.

688. What do the colors of flatworms signify? The black-and-white or the gray colors are those of the animal itself and probably have no particular significance, but the more vivid colors have a special interest. A flatworm, unless it is naturally so darkly pigmented that nothing else can show, takes on the color of what it has been feeding on. This is because its intestine branches out into all corners of the interior and is covered by only a thin layer of tissue, so that a flatworm feeding on purple sea squirts retains a purple color itself so long as it sticks to that diet; those feeding on orange sea squirts are a vivid orange themselves. The MacGinities, for instance, record that a thin and normally grayish flatworm was green when they found it crawling on blades of eelgrass, but that the same kind turned pink when fed with mussel eggs, only to become gray again three weeks later when the food had been fully digested.

689. How do flatworms reproduce? They are hermaphroditic animals, although self-fertilization is rare. Small flat patches of eggs are laid on the underside of rocks, from which free-swimming larvae hatch out.

690. Can flatworms repair injuries? Since their bodies are so soft and they crawl into hazardous cracks and corners, injury in nature is probably frequent. But of all animals the flatworms as a group are noted for their power of recovery or regeneration and have been employed

more than any other kind for scientific investigation into the nature
of regenerative growth. (See questions 171, 518, 699.)

691. How do flatworms move? They glide over the surface of the
rock or weed by means of invisible waves of muscular contraction
passing from head to tail. Some kinds, particularly the larger forms,
are able to swim by means of rapid undulations of their somewhat
frilly margin, so that the names "scarf dancer" and "skirt dancer"
have been given to such as these. (See question 251.)

692. Where are flatworms found? The smaller and less attractive-
looking kinds may be found under most rocks in the lower intertidal
region along every coast, clinging to the undersurface. There are many
kinds. The pale-yellow and light-brown species of *Leptoplana* are
examples common along the New England shore. A more striking
kind (*Stylochus*), brown with transverse yellow bands, and an inch
and a half long, is commonly found in dead whelk shells in southern
New England. On the Pacific Coast the flatworms come into their own
and are to be seen more in the open, with bright colors and lengths up
to six inches. Many are still unnamed. They may be found under
rocks, on rocks and beneath seaweed, or crawling over whatever they
may be inclined to feed on. Even their color is variable, depending on
the nature of their last meal. One particular Pacific Coast form
(*Planocera*) which is about an inch and a half long and almost as
wide, with beautiful blue-green, black, and white markings, is fairly
certain to be found in the Monterey region at least, if a certain amount
of effort is set forth. Large boulders on damp but not wet gravel
usually have several plastered to their undersides, and they are said
to richly reward any layman who undertakes to clear the surroundings
and roll over one of these huge boulders. The reward is esthetic, not
material.

693. What are ribbon worms? They are greatly elongated worms,
better known as the nemerteans, which are distinguishable from the
annelid worms (or true worms) by their lack of rings or segments,
and from flatworms by their great length relative to their width and
by the fact that their intestine is a straight tube passing from the
mouth at one end to the anus at the other. They are unique in having
an armed proboscis which can be shot out at great length from a spe-

cial sac at the front end of the body, to wound and wrap around their prey.

694. Where are ribbon worms found? The vast majority burrow in mud or sand or live on rocky shores amid rocks and seaweed or among mussel beds. The narrow forms (species of *Lineus*) are common in intertidal rocky regions, but the broader types (*Cerebratulus*) are generally found burrowing in sand or mud, at least by day.

695. How long can ribbon worms grow? As one biologist has put it, ribbon worms have acquired length but don't know where to stop. The longest is the so-called bootlace worm (*Lineus marinus*) which is a black worm about as wide as a narrow shoelace and often more than ninety feet long! On the American coasts, however, other species of the same genus exceed ten inches in length. On the other hand, the ribbon worm proper (*Cerebratulus lacteus*), the one most deserving of the name, may be an inch-wide flat ribbon anywhere from twenty to thirty-five feet long when expanded, although about three feet long and cylindrical when contracted. A West Coast species (*Cerebratulus herculeus*) found at Morro Bay is even longer.

696. Can ribbon worms swim? By night the giant forms (*Cerebratulus*) are inclined to swim up in the water, undulating their somewhat frilly margins as an aid to locomotion. (See questions 46, 722.)

697. What do ribbon worms feed on? Mostly annelid worms, which they capture by means of their proboscis.

698. How do ribbon worms find their prey? "When we see the nemertean *Paranemertes peregrina,* we can never resist the temptation to hunt up an annelid worm and draw it along ahead of the nemertean in the direction the latter is crawling. When the nemertean strikes this trail its proboscis shoots out and follows the trail to the annelid and then wraps around and around its body."—(G. E. and N. MacGinitie.)

699. What happens if ribbon worms are torn apart by crabs or other agencies? The most outstanding trait of ribbon worms is their tendency to break themselves into pieces when persistently disturbed.

This may be distracting to the individual but it can result in as many worms as there are pieces, since ribbon worms are remarkable for their capacity to regenerate and have been much studied for this reason. (See question 690.)

700. If a ribbon worm loses its long proboscis, can it feed? Not at once, but it soon grows a new one and then carries on as usual.

701. How small a piece of ribbon worm can regenerate into a new worm? At least in one of the eastern forms, there seems to be no lower limit provided the piece can heal, for which it must be twice as long as it is wide and must contain a fraction of the nerve cord. A worm may be cut and the pieces allowed to heal and then re-cut until miniature worms less than one hundred-thousandth the size of the original are obtained. In nature, of course, this could not happen, but it shows the possibilities.

702. How do ribbon worms breed? Apart from occasional fragmentation followed by regeneration, the separate sexes shed eggs and sperm into the water, where the free-swimming larvae develop.

703. How long can a ribbon worm live without food? A regenerating piece, or even an intact worm for that matter, can live for a year or more without food.

704. What are peanut and sausage worms? The names are given to two somewhat similar groups of unsegmented worms for which there is no really good common name. They are not seen by the casual observer and are accordingly not generally known. They are, however, abundant and interesting. One group, the so-called peanut worms, is called the Sipunculids and the other, the sausage worms, the Echiurids. When contracted they are tight and fully packed and have the general shape of peanut or sausage.

705. Where do peanut and sausage worms live? In burrows in mud flats, in sand, or in holes in rocks made by boring mollusks, from low intertidal levels into mainly shallow water. The peanut worms occur on both east and west coasts. On the East Coast the largest and commonest peanut worm (*Phascolosoma*), a foot long and one-third of

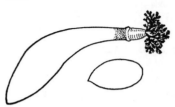

Peanut worm (1–3″), expanded and contracted

an inch in diameter, is abundant in shallow water along the entire
coast of New England, burrowing in mud or sandy mud, while a
small sausage type (*Thalassema*) lives in dead sand-dollar shells
along the North Carolina shore. On the West Coast the peanut worms
may be found in muddy crevices between rocks, while a remarkable
sausage worm (*Urechis*) known as the "fat innkeeper" burrows in the
mud at or below low-tide level in Newport Bay and Elkhorn Slough.
(See questions 707–712.)

706. How do peanut worms feed? A circle of tentacles around the
mouth is spread over the surface of the mud or rocks at the opening
of the burrow. Water is drawn toward the tentacles by the beating of
microscopic hairs, called cilia, which cover them, and the minute
organisms in the water are caught by slime and swallowed. When
disturbed, the peanut worm pulls in its crown of tentacles, telescoping
the head end completely within the body.

707. Which worm is called the innkeeper? The sausage worm
Urechis of Californian mud flats. The innkeeper is a cigar-shaped
worm, on an average about eight inches long but occasionally attaining nearly twenty inches. It has two golden bristles under the mouth
and a circle of bristles at the anal end.

708. Why is the innkeeper so called? Because in its burrow it is
host to several other creatures, namely a beautiful reddish scale worm
(*Harmothoë*), a pea crab (*Scleroplax granulata*), a small fish, the
common goby (*Clevelandia*).

709. How does the innkeeper make its burrow? It digs with its
short proboscis, scraping the walls of the burrow with its mouth
bristles. Then it crawls backward with its anal bristles until finally the

back end reaches the surface of the mud and the debris is blown out of the back door, so to speak.

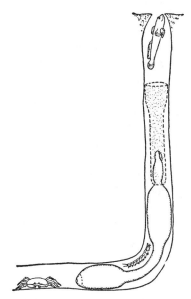

Innkeeper worm (4–5″) with fish, scale worm and pea crab as guests. Note slime net of innkeeper used for trapping microscopic food

710. How far apart are the two openings of the innkeeper's burrow? About thirty inches.

711. How does the innkeeper keep water flowing through the tube? By waves of muscular contraction which pass along the body of the worm.

712. How does the innkeeper feed? It spins a slime net with which it captures microscopic food. The net allows water to pass through but stops the smallest organisms. After a net several inches long has been formed and become clogged with food, the animal swallows the net and starts again.

713. What are annelid worms? Annelid worms get their name from the fact that their body is ringed externally—in other words, is annular. This obvious feature alone serves to distinguish the group from the various flatworms, ribbon worms, and round-worms.

714. What are bristle worms? The largest of the three divisions of the annelid worms, almost exclusively marine and to be found in all parts of the sea, from the mud flats and rocks along the shore to the depths of the ocean, while a few are found swimming freely in the water. They have a definite head, usually with eyes, and are notable in having bristle-like appendages along the sides of the body. They are more properly known as Polychaete worms, which means "many-bristled."

715. How do bristle worms live? In a great variety of ways, but principally as wandering predatory animals or as tube dwellers who trap and filter their food.

716. What senses do bristle worms possess? All have acute senses of touch and taste and most have one or more pairs of minute eyes. The more active the worm, the better developed as a rule are its eyes. Some that swim freely in the ocean have a pair of gravity-sensitive organs as well.

717. How do bristle worms move? The bristle-bearing appendages along each side of the body are usually equipped with both bristles and fleshy or paddle-like lobes. With these the worm can progress rapidly along the ground, burrow in sand or mud if need be, and swim to a greater or lesser extent according to how paddle-like the lobes are. The appendages are known as parapodia, meaning "lateral feet."

718. Can bristle worms repair injuries? Nearly but not quite all can readily replace any missing hind part, while a great many kinds can also replace lost sections in a forward direction as well.

719. How do bristle worms breed? In the very great majority of cases the sexes are separate and eggs and sperm are shed freely into the water, to develop into microscopic free-swimming larvae which drift with the currents for many days before settling on the sea floor.

720. Can bristle worms reproduce by means other than sexual? Some kinds do. Certain small and inconspicuous forms reproduce freely by forming buds or even chains of buds from the hinder part of the body, the buds breaking free to become independent individuals.

A few under certain conditions fragment into scores of short pieces, each of which regenerates a new head and tail and becomes a complete worm.

721. What worms are used as fish bait? Apart from angleworms or earthworms, which are neither marine nor polychaete but belong to another group of annelid worms, the worms commonly used as bait for fish are the lugworm (see question 782) and the so-called clam worm. (See questions 727–733.)

722. How does the moon influence the breeding of worms? Most marine worms spawn in tune with the tides and the amount of light reaching the sea, so that most are influenced by both the moon and the sun, at least to a marked degree. Two kinds, however, are affected to a remarkable extent, the palolo worm (*Eunice*) and the fire worm (*Odontosyllis*). The better known of the palolos is that of the South Pacific Islands, where the worm appears in immense swarms at the lowest tides along the reefs at sunrise during the last quarter of the moon during October and November. The spawning date is so regular that the natives know when to expect it and prepare for a feast accordingly. Through the West Indies and along the Florida Keys the Atlantic palolo is just as regular but its annual sexual orgy always takes place in July when the moon reaches its third quarter, although a preliminary spawning may occur at the first quarter. Fireworms, however, have a monthly date instead of an annual one, and during summer months at least spawn several nights in succession, commencing the third night after the moon is full. They must await the interval of darkness that comes after the sun has set and before the moon has risen, for they themselves are luminous, the females all over and the males with tiny headlights, in order to find one another in the water. Complete darkness is necessary for the lights to be seen. Swarms of flashing fireworms are thought to have been the lights Columbus saw on October 11, 1492, as he approached the outer reefs of the Bahamas. (See questions 46, 696.)

723. Which are the hunting worms? Many worms move about extensively in active search of living prey. There are many such, both large and small. The largest and best known forms are the beakthrowers (species of *Glycera*) and the clam worms (species of *Nereis*)

Beak-throwing worm (5–8"), with beak
or proboscis fully protruded

Head and jaws of clam worm (1/4")

common along both the East and West Coasts. Two others are of
particular interest because of their breeding habits. These are the
Atlantic palolo and the fireworm of the West Indies, occasionally
seen among the Florida Keys and northern Pacific Coast.

724. What is the beak-thrower and why is it so named? Beak-
throwers are whitish worms about six inches long, living in mud or
sand flats or among the roots of eelgrass where they prowl actively in
search of living prey. They are known as the beak-throwers because
of their ability to evert their sac-like proboscis out through the mouth
at great speed, turning the sac inside out. At the end of the sac are
four curved black hooks suitable for seizing prey or nipping an enemy.
When the worms are exposed, this organ is repeatedly extended and
withdrawn and is readily seen.

725. How do the beak-throwers feed? By shooting their long, wide-
mouthed proboscis at their prey. As the proboscis is drawn back the
hooks at the end close in. If the prey is a small crustacean or mollusk
it is then swallowed whole. The beak-thrower holds on and gradually
swallows and digests the whole as though it were a snake swallowing
another snake.

726. How do beak-throwers move about? When free-swimming,
which is frequent during the night, they coil themselves in a loose
spiral, rapidly rotating about their axis. When burrowing they insert
their pointed end in the sand or mud and rotate the body, to disappear
beneath the surface with great speed.

727. Where are clam worms found? Clam worms (*Nereis*) are wandering hunters like the beak-throwers (see questions 724–726) which live mainly in mud flats and in the soft muddy bottoms below low water. They also occur commonly under stones as well as in temporary burrows in sand or mud, in sheltered bays and inlets at low water and between the tides. The clam worm is not only one of the commonest but one of the largest of the marine annelids. It may reach a length of eighteen inches and is stout in proportion. The color is generally reddish, with iridescent green and bluish surface colors.

728. Can a clam worm hurt you? "The animals are very active and squirm violently when captured, protruding and withdrawing their chitinous jaws which terminate a wicked-looking protrusible pharynx and make carnivorous *Nereis* a formidable antagonist. These powerful jaws are capable of delivering a business-like bite to tender wrists and arms, but in collecting hundreds of them bare-handed we have rarely been bitten, always taking the precaution of not holding them too long."—(E. K. Ricketts and J. Calvin.)

729. How do clam worms feed? They have an eversible proboscis like that of the beak-throwers (see question 724)—except that it is smaller, more muscular, and is equipped with two curved black teeth instead of four—and it is used in the same manner. The worms feed on any small crustaceans, mollusks, and other worms they encounter, swallowing them whole. At the same time a clam worm can easily destroy worms and other creatures as large as or even much larger than itself. They do, however, set out actively in search of clams, and experiments show that no matter how well pieces of clam meat are hidden the worms search them out successfully, employing an obviously delicate sense of smell and taste. Not unexpectedly, therefore, clam worms are common in mud flats where clams are numerous. For a similar reason the worms are also common in mussel beds.

730. Can you distinguish the sex of a clam worm? "The male is easily distinguished from the female by its strong, steel-blue color, which blends into green at the base of the parapodia. . . . The female, on the other hand, is dull green, tinged with orange and red."

—(R. W. Miner.) This distinction is given for the common East Coast form but may not be exact for the Pacific Coast species.

731. What are the breeding habits of the clam worm? When clam worms are ready to breed, their appearance changes to such an extent that the sexually mature forms were originally considered to be a distinct kind of animal which was called the *Heteronereis*. The name has been retained, however, to denote the sexual phase. As the season approaches, the posterior segments swell up with either eggs or sperm, while the appendages normally used for creeping become changed into paddles for swimming. Then, after midnight in the dark of the moon during summer months, the two sexes leave the sea floor and swarm to the surface of the water. Ripe males are white where the sperm shows through the skin, while the back ends of the females are red with eggs. The eggs and sperm are shed into the water. When the process is over the worms drop back to their accustomed place and in due time grow new back ends since the old and exhausted sexual segments usually drop off. In a related East Coast form (*Nereis megalops*) with similar moon-breeding habits, common in Vineyard Sound, the female bites off from the male the posterior segments containing the sperm. Upon being swallowed the sperm go through the wall of the intestine into the body cavity to fertilize the eggs. The fertilized eggs later escape by rupture of the female's body, while in yet other species the fertilized eggs are incubated within mucous tubes in the mud by the male after the female has disintegrated during the process of spawning. (See question 722.)

732. Do clam worms have any enemies? Any animal that is numerous and actively moves about is in danger from more powerful creatures, and clam worms are no exception. When they leave their burrows—especially at night during the reproductive season, when they do so in large numbers to swim in regular undulating motion—they are readily destroyed by fish. Two kinds of fish, however, the tautog and the scup, nose them out of the sand to devour them.

733. Can any worms sting? No worm stings in the true sense of the word. Clam worms and beak-throwers can bite. But a certain pink bristle worm (*Eurythoë*) that lives in the mud flats along the Pacific Coast, such as those at Elkhorn Slough, has white glistening spines along each side. It is a striking-looking worm, and anyone who sees it for the first time will almost surely pick it up—a most regrettable thing to do. The spines readily puncture the skin and break off, with somewhat painful consequences. A larger West Indian representative of the same form is even more painful to handle and, in fact, is so immune from the attack of other creatures that it is the only wandering worm that feels free to roam at large in full view of the hungry world about it.

734. What are scale worms? For the most part they are small and rather flat worms an inch or so long which are clearly typically segmented worms when looked at from underneath, but whose character

Scale worm (1–1 1/2")

is masked on top by a double series of overlapping scales that give it an armored appearance. In one exceptional case, however, a giant form known as the sea mouse (see question 739), the scales are in turn disguised by a dense mat of hair.

735. Where can scale worms be found? Typically they are to be found in two places: clinging to the undersides of rocks in rock pools and the lower intertidal region generally (see question 5), commonly also among the holdfasts of kelp (species of *Polynoë* and *Lepidonotus*); and equally as commensals, that is, harmless companions, within the tubes of burrowing worms such as the innkeeper, the parchment worm and others (species of *Harmothoë*). (See questions 708, 764.)

736. What purpose do the scales serve? They probably possess both a protective and a camouflaging function, but which is the more important is not known.

737. Apart from their scales, do scale worms have any means of protection? The commensal forms obviously have found adequate protection within the tubes of the burrowing creatures such as certain worms and crustaceans. They have another which at least is good for the species however it may affect the individual. Whenever they are picked up, whether by a crab or a human, they usually wriggle so violently that they break into two, and under normal circumstances one at least of the two pieces has a good chance to live and regenerate the missing part. (See question 690.)

738. What do scale worms eat? They are predatory worms, though not averse to scavenging, and are equipped with a protrusible front end well supplied with horny teeth, feeding on any smaller creatures and fragments that come their way. (See questions 725, 728.)

739. What is a sea mouse? Possibly the most unwormlike worm there is, with roughly the size and shape of a mouse, although it is named for the Greek goddess Aphrodite! It lives in mud. Reported from the Atlantic Coast north of Cape Cod, it may possibly be encountered, particularly on the shore after storms. Large specimens may be six inches long, two inches at the widest part, and about one

Sea mouse (3–4″)

inch thick. The body tapers at both ends. "It is a scale worm, allied to the common and much smaller *Polynoë* of rocky shores, and bears fifteen pairs of large scales, or elytra, on the back. But these are entirely obscured in the intact animal by a dense covering of matted hairs, unusually long and thick bristles, which form a felting over the back. Laterally the animal bears other bristles, some stout and used in locomotion and others, probably protective, which shine with a brilliant iridescence, with golds and reds and other colours, and give to this obscure inhabitant of the sea bottom a quite surprising beauty."
—(C. M. Yonge.)

740. How does the sea mouse live? "The sea mouse lives beneath the surface with only the hind end exposed. By appropriate body movements water is drawn in along the underside and then ascends in streams between the laterally projecting parapodia and so into the respiratory space between the back of the animal and the over-arching scales. The latter are then drawn down and the water expelled by way of the upper side of the tail. The thick felting above and the lateral iridescent bristles protect this respiratory current from contamination or blockage by the surrounding muddy sand. The internal arrangements are no less striking. There is an immense muscular region behind the mouth—opening when the carrion which the animal encounters and swallows in its slow movements is ground up."—(C. M. Yonge.)

741. Which worms build tubes of lime? A group of worms known as the serpulids because of their sinuous tubes of lime, the tubes being either cemented to the surface of rocks or weeds, or in tangled masses with other tubes, or deep in the substance of coral with only the ends exposed.

742. Where are limy-tube worms found? A beautiful solitary worm (*Hydroides*), about three inches long, builds its long, twisted white tubes in mollusk shells, particularly on the inner surfaces of empty ones, and is very common from Cape Cod to Florida. The great twisted masses of a similar worm (*Serpula*) cover the rock reefs of the Puget Sound to Strait of Georgia region of the Pacific Coast, and in less abundance the worm may be seen attached to the sides of rocks as far south as Monterey. It also is found in shallow water on the

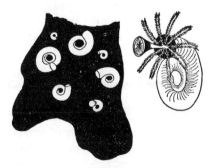

Spiral limy-tube worm, showing tentacles and tube plug (1/8")

east coast and Europe. The tangled masses of a much smaller kind (*Filograna*), with long but very slender white tubes, is common along most of the Atlantic Coast, attached to the sides and undersides of rocks together with encrusting sponges. While last but not least, because they are so abundant, the undersurfaces of rocks and the blades of kelp and lower intertidal seaweeds are generally covered by the minute white coiled tubes, no more than one eighth of an inch across the coil, of the smallest serpulid of all (*Spirorbis*), along both the Atlantic and Pacific shores.

743. How can limy-tube worms best be seen? When disturbed in any way the worms are tightly locked within their tubes and only the tubes are seen. To be fully appreciated for their true wonder they should be kept in a dish of cool sea water. Then the heads will protrude and, preferably with a hand lens, the magnificent color and pattern of the crown of tentacles and the operculum or plug can well be seen.

744. How do limy-tube worms make their tubes? They precipitate calcium carbonate from the sea water as a ring of lime in a fold of the skin (the collar) just behind the tentacles. This is added to whatever has already been laid down, so that the tube is slowly extended forward as the worm grows.

745. What are the colors for on limy-tube worm tentacles? The rich bands of color give the tentacle crown a beautiful appearance, for the crown may be purple with white and green bands, or brown with white and yellow, or brilliant red (in *Serpula*). So far as is known, the pigment serves no particular purpose and may be simply the result of the fact that the worm is safe in its tube and can display color with impunity. Small eye spots on the tentacles enable the worm to withdraw to safety instantaneously, and it is apparently in no danger from other animals. (See question 688, 757.)

746. How do limy-tube worms feed? By protruding a crown of tentacles consisting of two semicircular fans which draw water and its contained food particles in toward the mouth of the worm. Seen under the hand lens, the crown is one of the beautiful sights of marine life. The tentacle crown is very similar to that of the feather-duster worms and works in the same way. (See question 756.)

747. How do limy-tube worms close their tubes against intruders or against loss of water when exposed between the tides? One of the tentacle filaments has a large cone-shaped knob, usually of striking color pattern, which plugs the opening of the tube as the worm withdraws. It is called the operculum.

748. If the tube plug (or operculum) is torn off, is it replaced? Yes, but not from the stump of the old one. A new one grows from a very short filament on the other side of the head, which seems to be held in reserve just for this contingency. (See questions 545, 601.)

749. How do limy-tube worms reproduce? The larger kinds are separately sexed and shed eggs and sperm into the sea water during the breeding season to become free-swimming larvae. The small coiled *Spirorbis* however is hermaphrodite, the front part being female and the hind part male; the embryos develop within the tubes of the parent and are liberated ready to settle at once on the rocks and weed. The small clustered worm *Filograna,* however, not only reproduces sexually but also by producing buds from the hinder end. (See question 742.)

750. What worms live in mud tubes under rocks? They are tube worms fairly closely related to the feather-dusters though of a kind which move about to some extent and renew their soft tubes of mud, which are in any case little more than the consolidated walls of a burrow. They have tapering, flesh-colored or pink bodies, three to twelve or more inches long, which bear a writhing mass of sinuous tentacles at the front end. They are commonly found under rocks at low tidal levels where the rock is well bedded in some mud, one form in particular (species of *Amphitrite*) ranges along both the Atlantic and Pacific Coasts.

751. What is the ornate worm? Similar to the mud-tube worms that live beneath the rocks except that it builds better tubes and lives more in the open, and is a species of the same genus (*Amphitrite ornata*). "This is one of the most beautiful flower-like worms along the coast. Its tubes have a round opening, one-quarter inch or more in diameter.

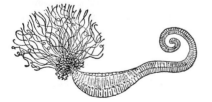

Ornate mud-tube worm (3–7")

When built in sandy regions, it is surrounded by a low mound of sand, often widely different in color from that in the surrounding sea bottom. It is common around and below the low-water mark, in sand and gravel, from Vineyard Sound to the New Jersey Coast."—(R. W. Miner.)

752. How are the mud-tube or sand-tube worms best observed? Put one in a small vessel of sea water and watch for awhile. "A specimen of any of these worms, deprived of its home and placed in an aquarium with bits of gravel, will quickly set about remedying the deficiency. In a few hours a makeshift burrow will have been constructed of the gravel, cemented together with mucous and detritus, and the transparent stringy tentacles will be fully extended on the bottom of the container, writhing slowly in search of food, which is transferred to the mouth by means of cilia."—(E. K. Ricketts and J. Calvin.)

753. What are the tentacles for at the head of the mud-tube worm? The tentacles are of two kinds. There are three pairs of blood-red branching gills which are comparatively short, and numerous extensible flesh-colored tentacles which are in constant motion and serve for feeding. The feeding tentacles when extended reach out for relatively enormous distances from the head in all directions. They operate in a manner all their own, for they serve as conveyor belts, for the minute hair-like cilia which cover them beat so as to pass back to the mouth any microscopic organisms that come in contact with them.

754. What are feather-duster worms? Feather-duster worms, also known as peacock worms, flowering worms, or plumed worms, are worms that live in long, narrow parchment tubes of their own making,

Feather duster worms (6–18″), with crown of tentacles protruding from parchment tubes

with at least half and usually more of the tube standing free in the water and the flower-like head extending from the end of the tube. Many of them rank among the most beautiful and striking creatures in the sea. They are known zoölogically as the sabellid worms.

755. Where are feather-duster worms found? Worms up to eight or nine inches long (*Sabella*) living in tubes about one quarter-inch wide are common among the Florida Keys and Gulf Coast, either partly buried in muddy sand at extreme low tide or attached in large numbers to mangrove roots below the water line. Another, the slime-tube duster (*Myxicola*) lives in mucous tubes buried in sandy mud along the Gulf Coast flush with the surface. A dwarf species (*Potamilla*) may be found along the Atlantic Coast as far north as Cape Cod attached to the strings of empty egg cases of the channeled and knobbed whelks. The outstanding Pacific Coast form is a sabellid giant (*Eudistylia*) which is one of the vivid sights of North Pacific tide pools and wharf piles, from Alaska to Monterey; the worm itself may be as thick as your little finger and a foot and a half long, while its large plume of tentacles is a rich maroon.

756. What are the plumes or feathers of the feather-duster worms for? Primarily for feeding, though also for respiration. They correspond to the pair of head tentacles most marine worms possess, but here each member of the pair is subdivided into a considerable number of long filaments each of which bears innumerable small side branches just like a feather. Altogether they constitute a funnel, and the ciliary hairs beating on all the branches cause a current of water to converge on to the mouth, bearing minute food organisms. (See question 746.)

757. Why are the feather-duster tentacles so highly colored? The dark red, brown or purple bands, and dark pigment generally, on the tentacles are for the most part light-sensitive, enabling the worm to sense when the tentacles are fully protruded and also to detect any sudden changes in light and shade which may spell danger. (See question 745.)

758. How can a feather-duster be distinguished from a sea anemone? As a rule there is little chance for a mistake. But the large

maroon duster of the Pacific Coast looks remarkably like an anemone
when fully protruded, and so does the slime-tube duster of the Gulf
of Mexico. The distinction however is obvious when they are disturbed, for the worm heads disappear instantaneously if a shadow is
passed over them, while anemones withdraw fairly slowly and are
more likely to respond to a sudden light than a passing shadow.

759. How do feather-duster worms protect themselves? A feather-
duster with its colorful head protruding is obviously a tempting morsel
to any passing fish, and feather-dusters clearly have the uncomfortable
choice of feeding dangerously or starving in safety. Consequently they
feed when all is quiet and shoot rapidly down to the lower part of the
tube at the slightest disturbance. In particular they respond very
quickly to changes in light intensity such as those created by a passing
shadow. All have the so-called shadow reflex to a greater or lesser
degree. Those with the fastest response generally have small eyes at
the tips of all the numerous tentacles and are rarely caught with their
necks out. Those without eyes but with bands of sensitive pigment
giving pattern and color to the tentacles have a slower reflex, and
examination shows that about half the individuals in a community
have lost their heads and grown new ones. In this type a sudden pull
on the tentacles leaves whoever has done the pulling with only the
crown as booty, for the worm throws off the crown with a snap and
quickly grows another. (See questions 545, 774.)

760. How do feather-duster worms make their tubes? The inner lining of tough integument is secreted from the outer surface of the
worm itself from special glands. The material produced toughens into
a substance much like that of the parchment worm, but the sand
grains or other particles which cover the outside of the tube and form
an integral part of it are brought up from below and are passed along
the whole length of the worm to the region of the head, where the
collar fold of the worm plasters them into place. (See question 744.)

761. Can worms build reefs? The honeycomb worm (*Sabellaria*)
builds reefs of a kind, although the reefs do not have the hardness of
coral reefs. It often forms massive reefs, however, on the exposed surfaces of rocks where these are more or less bounded by sand at low
tide. The masses are frequently dome-shaped. (See question 292.)

762. How do honeycomb worms form a reef? Coarse sand is cemented together with sticky mucus by the worm's lip or building organ to form a tube. The tube of one worm, however, becomes cemented to the next, as well as to any rock surface it may touch, until tubes overlap so as to form large, thick, encrusted areas, with only the open ends of the tubes exposed. The worm reefs have somewhat the consistency of porous sandstone.

763. Where can honeycomb worms be found? They are very common from Vineyard Sound to New Jersey and at various places along the Pacific Coast, more particularly between San Francisco and San Diego. Honeycomb reefs are equally common along the coasts of northern Europe.

764. What is the parchment worm? It is a tube worm (*Chaetopterus*) notable both for its tube and its anatomy. The worm itself is from six to fifteen inches long and lives within a U-shaped tube of parch-

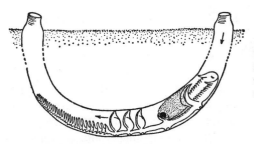

Parchment tube worm (5–8")
in imbedded tube, showing fans
for water current and slime
net for trapping food

ment of its own making, the two ends of the tube extending above the surface of the ground. The projecting ends of the tube are narrow compared with the submerged part.

765. Where are parchment worms found? Buried, except for the short protruding ends of the tubes, in mixtures of sand, mud and gravel from low-tide level down into moderately deep water. It ranges from Cape Cod to North Carolina on the Atlantic Coast, Newport Bay to Vancouver on the Pacific, and along the coast of northern Europe. They also occur along the Gulf Coast of Florida.

766. How can you recognize the tube of a parchment worm? The parchment nature of the tube is very evident. The material is like very tough moist paper wound in several layers. Empty tubes are often tossed up by the tide. Inhabited tubes still buried in the sandy mud at or just below the low-tide level can be identified by the two chimneys usually light brown or dirty cream in color, extending about an inch above the surface.

767. Is it always dark within the tube? Apart from such stray light as may enter the openings of the tube, the interior is illuminated by the animal's own phosphorescence. (See question 101.)

768. What is the reason for the parchment worm's phosphorescence? No good purpose is at all evident. The light, however, may repel some kinds of undesirable intruders and may equally well attract various small organisms of a size and kind suitable for food. The phosphorescence is associated with the large quantities of slime pro-

Parchment tube worm as seen in dark, showing phosphorescence (6–8″)

duced by the worm and may even be an accidental but harmless quality of the slime which is simply tolerated but not made use of. The slime covers the body of the worm and lines its tube. When the worm is touched, some of the slime sticks to your fingers and continues to glow there.

769. How are parchment tubes made? "The larva soon ceases to swim and, becoming sluggish, settles to the bottom. It begins to creep about, perhaps finding food in the diatoms lying in the troughs of the sand ripples. As it creeps it leaves a trail of mucus. After perhaps a few days the young worm begins to make short, mucus-coated tunnels, burrowing into thick clumps of diatom mixed with sand. From such a simple tunnel, extending perhaps several times the length of its body, the larva pushes up extensions to the surface of the sand, to create the U-shape. All later tunnels are the result of repeated remodellings and extensions of this one, to accommodate the growing body of the worm."—(Rachel Carson.) (See questions 744, 760.)

770. How big is the tube of the parchment worm? A fair-sized tube is about one inch in diameter at its widest part and close to two feet long.

771. How does the parchment worm hold its position within the tube? The water current is so powerful that without some anchoring device the worm itself might be swept backward or be carried forward by its own pumping movements. A series of four suckers along the ventral side of the worm at the fans and the neck accordingly hold the animal firmly in position while the fans are in regular motion. (See question 628.)

772. How does a parchment worm create the strong current of water flowing through its tube? Three of the body segments, about midway along the length of the body, are specially modified as the so-called fans. Each of the three is a wide, nearly circular, rhythmically beating fan which practically fills the diameter of the tube. The three beat in regular succession, acting as a pump, and cause an astonishingly strong current to flow through the long tube. (See questions 711, 784.)

773. How does the parchment worm feed? "On the dorsal side of the worms are the structures concerned with feeding. About the middle of the body there are three 'fans' that are used to propel the water through the body. Near the head end are two arms or wing-like structures that are held tightly against the inner circumference of the tube. Thus the body of the worm with these two 'wings' forms a circular shelf extending inward from the walls of the tube, leaving a hole in the middle of the shelf. Mucus is secreted from the inner edge of these 'wings' and is pulled out into a long bag by cilia lining a groove between the 'wings' but ahead of the 'fans.' Here the lower end of the mucous bag is gathered into a small pellet. As the worm feeds, mucus is continually secreted by the 'wings,' and the other end of the mucous bag is rolled up by the ciliated 'cup' until a pellet the size of a BB shot, or smaller, is made. Then the worm stops secreting mucus, reverses the movement of the cilia in the groove, and thus rolls the pellet up to the mouth where it is swallowed. Since all water has to pass through the mucous bag, the pellet contains every particle of material that entered the burrow with the current of water while the bag was in place." (MacGinities) (See questions 712, 781, 787.)

774. Does the parchment worm ever leave its tube? Not intentionally. Its large, creamy-white body would be snapped up by any passing crab or fish. Yet there is evidence that it not infrequently sticks its head out, since many specimens are found within their tubes without their original heads, and with small new heads growing in place of the old. (See questions 545, 759.)

775. What are mason-worm tubes like? The tubes are beautifully made, graceful, trumpet-shaped cones, about two inches long, constructed of sand grains which are small at the narrow end of the tube and larger at the large end. There is an opening at both ends. The sand grains are so carefully fitted together that the tube is almost smooth externally. Mason worms are sometimes called trumpet worms.

776. Where do mason worms live? They live in sand or sandy mud, mainly below the low-tide level, although they are sometimes found in sandy pools in the low tidal zone. Their empty tubes, washed up after storms, are more commonly seen along the shore. The East Coast mason (*Cistenides*) is abundant from Maine to North Carolina, the West Coast mason (*Pectinaria*) occurs in Elkhorn Slough and at various other localities. Unlike most tube worms, the mason worms carry their tubes around, working obliquely into the sand head downward with the narrow end of the tube projecting for a short distance into the water above. After they have fed in one spot for several hours, they move along to another.

777. How are mason-worm tubes made? According to A. T. Watson, "Each tube is the life work of the tenant and is most beautifully built with grains of sand placed in position with all the skill and accuracy of a human builder." The finished wall is only one grain thick and the grains are fitted so that projecting corners fill the angles and exactly the right amount of cement is used. The sand grains are carefully selected by the mouth and cemented into position "after which the worm applies its ventral shields to the newly formed wallings, and rubs them up and down four or five times, apparently to make all smooth inside the tube. The moment when an exact fit has been obtained is evidently ascertained by an exquisite sense of touch." (See questions 744, 752, 760.)

778. What is the appearance of the mason worm itself? The most outstanding feature is the great golden combs used for digging. The body is flesh-colored and tapering, usually mottled with dark red and blue. Two pairs of bright red gills at either side of the head are also conspicuous.

779. How do mason worms burrow? Digging is a vigorous process and is carried out by means of a pair of prominent, golden bristle-combs which project from the head. Sand is tossed by left and right strokes alternately. (See question 786.)

780. How do mason worms breathe? A respiratory current of water is drawn in through the narrow end of the tube in spite of sand being passed out of the same opening. The opening can, however, be plugged by a fold of the body if there is any danger of sand entering and blocking it. There are two pairs of red plume gills at each side of the head for respiratory exchange.

781. How does a mason worm feed? It feeds largely on organic debris from near the surface, but it obtains the debris in a rather re-markable manner. It makes a funnel-like opening through the sand

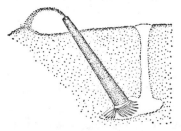

Mason-tube worm (2") burrowing and feeding, and removing sand

to one side of its tube. As it digs away, head down, passing sand up its tube to fall on to the surface, material keeps falling down the funnel and brings with it a new supply of food. (See question 787.)

782. What is the lugworm? Lugworms are fat, greenish-black worms three to twelve inches long which are extensively used as bait by

Lug worm (6–18″) showing large
mouth and pairs of skin gills

fishermen on both sides of the Atlantic and on the Pacific Coast as well. They may have been so called because of the effort required to collect them from their burrows, or more likely perhaps because of their sluggish or laggard nature. Their scientific name, *Arenicola,* means "sand-dweller."

783. Where can lugworms be found? They burrow in sand just below the low-water mark along the entire coast north of Long Island Sound, and also along the shores of Europe. On the Pacific Coast they are abundant in the low-tide mud flats in Puget Sound but also occur, though less commonly, farther south.

784. What kind of burrow does a lugworm make? Usually the burrow is U-shaped and is of a size that permits the greatly swollen head and the tail of the worm to reach the surface. A current of water is created through the burrow by muscular waves of dilation produced along the body. The current can be made to flow in either direction, and unlike the parchment worm, without having to turn around. (See questions 772, 711.)

785. How can lugworm burrows be recognized? They make castings of undigested sand or mud at the surface at one end of the burrow. Under water these are washed away probably within a matter of minutes, but if formed after the tide has exposed the beach or mud flats where the worms live, the castings remain until the tide returns and are an unmistakable sign of the lugworm's presence. The castings look just like those of earthworms. The only marine animal that makes similar castings is the so-called acorn worm, which is not a worm. (See question 905.)

786. How does a lugworm burrow? In very much the same manner as when feeding, except that the bulbous proboscis is used mainly for forcing the sand or mud aside in all directions. When burrowing primarily, as distinct from feeding, very little mucous is produced on the proboscis, so that material does not stick to it.

787. How do lugworms feed? The worms have a large proboscis which can be everted like a funnel. The outside of this organ is kept sticky with mucus to which sand, and any material mixed with it, adheres. When the proboscis is pulled in the trapped material is swallowed. As the worm moves forward while it makes its burrow, much of the material ahead of it consequently passes through its intestine, and most of it will be indigestible and will be finally evacuated in a cast (see question 785). An actively feeding worm makes castings about every half hour. The worm swallows about a dozen times in succession at about five-second intervals, rests a few minutes, and then begins again. The animal lives on whatever organic matter or small organisms may be contained in what it swallows. (See question 773.)

COELENTERATES

788. How do sea anemones feed? All are carnivorous and are designed to trap living prey. In all cases the tentacles are equipped with sting cells, and the paralyzing power of the stings and the retractile nature of the tentacles enable the anemone to bring to its mouth most

Dahlia anemones (1–2"), contracted and expanded

small creatures which encounter it. Anemones with large tentacles which are not excessively numerous depend upon small shrimps, crabs and fish. The plumose anemone on the other hand, with a thousand or more very delicate tentacles, relies mainly upon the much smaller swimming crustaceans and other forms, and also upon the microscopic food brought to its mouth by the currents of water the tentacles create.

Others again, like the burrowing *Cerianthus,* are effective traps of an intermediate sort. This anemone, for instance has a column about three quarters of an inch wide, but its tentacles, spread out over the sand, cover an area two feet in diameter or about 450 square inches. Any small animals inadvertently crawling across them are promptly stung and drawn to the mouth. (See question 802.)

789. How fast can an anemone digest its food? In the case of the plumose anemone "a small chiton (a mollusk) which had been cut in two was placed on the disc of an anemone in a glass dish. The disc was immediately depressed at the center, so that the food disappeared into the body cavity. Within fifteen minutes the cleaned shell of the chiton, all the meat completely digested away, was disgorged through the mouth—a striking tribute to the potency of *Metridium*'s enzymes."—(E. K. Ricketts and J. Calvin.)

790. What animals feed on anemones? Some fish, such as cod, feed on anemones if opportunity arises, but the only creatures which are specialists in the matter are certain kinds of sea slugs. These are apparently immune to the stings of anemones and eat them with impunity. Most other creatures that tackle an anemone larger than themselves are likely to be consumed, for even the armor of a coat is inadequate protection. Surprisingly, too, the sting cells of either anemones or hydroids that have been eaten by sea slugs remain alive and somehow work their way through the sea-slug tissue to lie at the surface, thereby giving added protection to the slug. (See questions 364–366.)

791. What meaning do the colors of anemones have? Anemone colors and pattern are mainly for camouflage, though for offense rather than defense, serving to disguise the anemone from its prey rather than its generally non-existent enemies. Likewise the flamboyant color of some anemones may be a deluding device or more likely a careless exhibition of color by creatures that have little to fear from advertising their presence.

792. Can sea anemones move? Some never do, but most of them can, slowly but surely. Plumose anemones that find themselves unexpectedly exposed to too much light move to a more shaded position within a short time. Three or four inches an hour is the record, but even that seems remarkable for a tree-like column with a base not much wider. The flat base does the work, slowly gliding by means of imperceptible muscular waves that pass along it. *Sargartia* (see question 796), when brought to an aquarium attached to sea lettuce or some other seaweed, quickly leave the plant and attach themselves to the smooth glass sides of the aquarium.

793. What senses do sea anemones have? They have no sense organs or any sort of brain, but the network of nerves underlying the skin is sensitive to light, touch and some chemical substances. It is enough to serve their way of life.

794. How large can an anemone be? The solitary great green anemone of the Pacific Coast reaches a diameter of about ten inches. A plumose anemone dredged from sixty fathoms off the same coast filled a ten-gallon crock when expanded. The record is reached, however, by giant anemones of the Great Barrier Reef of Australia and of other Southwest Pacific regions, which attain a diameter of over three feet, forming chasms capable of digesting almost anything.

795. How long can anemones live? Individual anemones have been seen to occupy the same crevice of a rock pool for more than thirty years. Furthermore, the University of Edinburgh in 1900 received some anemones from a woman who had collected them, already fully grown, thirty years before. They were still in good health forty years later and may yet be alive, aiming at a century. On the other hand, plumose anemones may reach a fair size within three or four years, but unless they live in very quiet waters or are anchored deep within a crevice, they are all too likely to lose their hold on the rock and come to an untimely end. Under proper circumstances there seems to be no reason or evidence that anemones ever die of old age.

796. Can anemones grow from small pieces? More readily than can any other kind of animal. The capacity is exploited as a regular means of reproduction by a number of species. The process is called pedal laceration and is commonplace among the plumose and *Sargartia* anemones and certain others. According to the MacGinities, the base of a *Sargartia* spreads out on the glass of an aquarium and becomes firmly attached at its outer edge. The central portion then pulls away, leaving behind it a more or less complete ring of pieces, each of which in turn may remain whole or split up into two or three tiny pieces, resulting in a dozen or more offspring. Each one of these pieces then develops tentacles and a mouth and becomes a full-fledged though minute individual. The separating of the pieces from the parent individual is a slow process, requiring from several hours to one or two days. One specimen that they watched gave rise to a dozen buds

Plumose anemone (3–7″), with propagative fragments left by contracting base

in less than four hours and to about thirty in the course of three weeks. The ability to bud off new individuals in this manner is dependent upon the food supply, for after budding offspring, the parent proceeds to feed and store up material, and grows perceptibly in size. Then again it buds off another group of offspring. When the parent was well fed, buds were given off every three or four days, but a starving animal did not bud at all. (See questions 699, 749.)

797. How do anemones propagate? By one or more of three ways: by simple division or fission, by budding, and by the sexual method. Simple division, or direct splitting into two individuals, is comparatively rare although in the case of the green anemone carpeting the California rocks it is a regular process. This anemone takes about two days to complete the process of pulling apart. As long as individuals of a cluster are pressed tightly together, nothing happens. But any individual on the outer edge of the group, or placed in a solitary position, seemingly feels its freedom and divides into two, the split starting at the tentacle end and proceeding steadily toward the base. By this means a single anemone can form a sizable colony at the end of two or three years. *Sargartia* anemones often divide in the same way. Budding off fragments of the foot, or pedal laceration, is typical of plumose anemones and *Sargartias* and is described under question 796. Yet another kind (*Boloceroides*) constricts off its tentacles at their base and a new anemone then grows from the severed end of each detached tentacle. All such means as these serve to increase the population of anemones in particular localities.

Anemone (Epiactis) (1")
with young in brood pockets

Plumose anemone dividing
into two from top downward

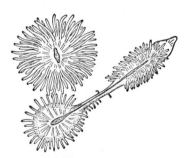

Pacific green anemone (1") dividing
by means of extension of base

Sexual reproduction, however, is universal among anemones whether splitting or budding processes occur or not, and individuals are always of separate sex. In most cases, as in the plumose anemone, *Metridium,* sperm and eggs are discharged from the body cavity through the mouth, and chance unions of these products carry the species far and wide throughout the Northern Hemisphere. After a week or so of development and gliding through the water, the larvae settle down on the sea floor or along its margin and those that are lucky survive and grow. In the case of certain species, however, the larvae become temporary parasites on jellyfish and comb jellies; during this time they grow markedly and are also carried over vast distances by their hosts. When they finally sink, usually when the jellyfish disintegrates, they are no longer larvae, but small well formed anemones relatively capable of establishing themselves. In other cases, such as the small anemone, *Epiactis prolifera,* which is locally abundant on the Pacific Coast on the protected sides of small smooth boulders, the eggs, instead of being discharged into the water, migrate out through the mouth and down the sides to occupy brood pits at the base of the parental column, where they develop. A fully expanded adult, with half a dozen flower-like young, is a lovely thing to see. As they reach a larger size they crawl away and become free. (See questions 807, 810, 822, 863, 873.)

798. What are the nearest relatives anemones possess? The hard corals. Soft corals are not so closely related.

799. Where are sea anemones to be found? Along the shore anemones are most likely to be found anchored in the crevices of rock pools at low tide, under protected over-hanging ledges where the tide rarely leaves them, attached to wharf piles below the low-water line, under floats, particularly those which are left in the water from one year to the next, and in the case of certain especially adapted kinds, more or less buried in mud flats or muddy sand. Several are outstanding for either conspicuousness or interest. (See questions 800, 801, 802, 803.)

800. Which anemones live on rocks and wharf piles? The plumose anemone (*Metridium*) is undoubtedly the best known, being common on both sides of the Atlantic and also on the Pacific Coast. It is one of the handsomest as well as the commonest kinds and occurs on wharf piles, rock pools and in rock crevices generally. When contracted, as they always are when the sea withdraws from them or when they are otherwise disturbed, they look like a brownish or yellowish blob of stiff opaque jelly, about as unbeautiful as it could be. When expanded, the velvety smooth column, which may be olive brown, white and brown marbled, white, orange, or peach color, expands up to six or eight inches in shallow-water individuals, while the crown opens to show as many as a thousand slender tentacles attached to a wavy base. In spite of appearances the anemone can move slowly over rocks, etc., if uncomfortable where it is, and in the process usually leaves fragments of its attached area behind. Or it may contract its base as a whole, leaving a ring of fragments of irregular size around itself. In either case the fragments grow tentacles and form a mouth eventually growing into full-sized individuals. (See questions 796, 797.)

801. Which anemones live in rock pools? On the Pacific Coast two green anemones are common, both of them forms of *Bunodactis* alternately known as *Cribrina*. Both are green for the reason that their tissues are loaded with microscopic green algae, an association which works to the metabolic advantage of each partner. Accordingly these are light-seeking anemones, for otherwise their guest algae could not function well, and one of them, the solitary great green anemone, is perhaps the most obvious inhabitant of the lower tide pools between Alaska and Panama. Anemones with a ten-inch spread are not uncommon. The other form is very likely a special size and color phase of the large solitary one, occurring on the exposed surfaces of rocks, especially where sand is being deposited, packed tightly together to form extensive carpets. "It is the beds of these hardy animals, certainly the most abundant anemone on the coast, that makes the intertidal rocks so 'squshy' under foot, the pressure of a step causing myriads of jets of water to extrude from the mouth and pores of the body wall. These anemones commonly cover themselves with bits of gravel or shell, so that when contracted they are part of the background. Many a weary collector has rested upon a bit of innocent-looking rock ledge, only to discover that an inch or so of exceedingly and very contagious wetness separated him from the bare rock."—(E. K. Ricketts and J. Calvin.)

Also notable for its ornamental beauty is an East Coast form, the dahlia anemone (*Tealia* or *Urticina*), which is especially abundant from Nantucket to Block Island, with its red-crimson blotched column and striped thick tentacles. (It is not unknown on the Pacific Coast but is less likely to be seen.) Smaller anemones (species of *Sargartia*) with long slender columns attached to buried stones on the sea bottom or to the rock beds of partly sand-filled rock pools are also common in the same general region, from Cape Cod to Long Island Sound. Small *Sargartias* are common on the Pacific Coast. (See question 5.)

802. Which anemones burrow in sand? While anemones as a whole are primarily adapted for living attached to hard rock or similar surfaces, some have managed to take advantage of more wide-open spaces and have become dwellers or burrowers in mud or firm sand. On the West Coast the slender *Edwardsiella*, one or two inches long, occurs in great profusion. As many as fifty per square foot crowd together in sandy mud banks near Corona del Mar, each worm-like body, almost covered with a brown and wrinkled tube, protruding slightly above the surface. Another burrowing anemone, common in Southern California is *Harenactis*. "It is a tubeless form, sandy gray in color. The long wrinkled worm-like body is largest in diameter at the disc, tapering downward until it swells into an anchoring bulb. When annoyed, *Harenactis* pulls in its tenacles in short order, retracts into the sand, and apparently sends all its reserve body fluids into this bulb, thus swelling it out so that it is impossible for a hungry bird to pull the animal out. The bulb of a large specimen may be eighteen inches below the surface so that digging out a specimen with a spade is very difficult, as the shifting sand fills up the hole almost as rapidly as it is made. While the excavation is in progress the anemone retracts still more and makes its bulb even more turgid. Having done that, it can do nothing but sit tight, a fatalistic philosophy that is in this case usually justified." (E. K. Ricketts and J. Calvin.)

Another burrower, *Cerianthus,* with flower-like head, lives in large numbers in sandy shoals and in the floor of sandy creeks from the Carolinas to the Gulf of Mexico, particularly among mangrove inlets. As you reach for the chocolate-brown heads they disappear as though they had been an illusion. They are remarkable in that they use their sting cells to bind their mucous or slime tube together, a surprising use for an offensive weapon.

803. How else can an anemone live? In contrast to both the sand burrowers and the stand-pat rock dwellers, an anemone with the suggestive name *Adamsia tricolor,* lives attached to the shells inhabited by hermit crabs, traveling hither and yon without any effort of their own. (See question 633.)

804. What are hydroids? Small polyps, like miniature sea anemones, which produce buds which fail to separate from the parental stock

Hydroids (1/2–1″)

and so give rise to colonies growing either as a cluster from a common base or as a delicate feathery, tree-like growth, attached to kelp, rocks and wharf pilings along the coasts.

805. How do hydroids live? Much as do sea anemones, each small polyp of a hydroid being equipped with a ring of tentacles around the mouth, with which it captures small swimming crustaceans and other organisms of food. The tentacles are supplied with stinging cells of the same general kind as those of anemones and jellyfish and are employed to kill or immobilize the prey. (See questions 788, 816.)

806. Is a cluster or colony of hydroids one or many? It is one colony made up of numerous conjoined individual polyps. What constitutes the individual depends on how you look at it. (See questions 836, 846.)

807. How do hydroids propagate? Apart from their peculiar form of growth by budding and branching, which gives rise to progressively larger colonies but rarely to increase in colony number, hydroids reproduce in one of two ways. Either they produce special buds which break away and swim off as small jellyfish, or they produce eggs and larvae which settle in the neighborhood of the parental colony, thereby increasing the local population. The small jellyfish drift with the currents and become sexually mature when still small, usually when from one quarter inch to one inch across. These in the main are the thimble jellies, and their eggs and larvae in turn settle on the sea floor to develop into hydroids, though perhaps scores of miles from where the parental colony existed. (See questions 797, 819, 847.)

808. Do all hydroids give rise to jellyfish? Some do and some do not. Those that do are for the most part small, inconspicuous colonies not often seen, while the larger and more obvious hydroids found in pools on rock weed and wharf piling, with one or two exceptions, do not.

809. What hydroids are of particular interest? Half a dozen hydroids are comparatively spectacular and interesting. The most conspicuous are the reddish flower-like clusters of *Tubularia* which are common on wharf piles and floats, each white or yellow stalk of a cluster being about two inches long and bearing a reddish polyp at its end. A hand lens shows grape-like branches hanging down among the tentacles, which are the reproductive bodies. Active larvae emerge from them and settle down in the immediate vicinity to give rise to new colonies. The rocks at low tide and in shaded, low-level pools along the New England coast are regions where *Tubularia* colonies often abound. Somewhat similar though somewhat smaller, colonies of the pink or reddish, white-stalked hydroid *Clava* are frequently found commonly attached to knotted-wrack and bladder-wrack seaweeds and under

Hydroid (1/8″) covering shell
inhabited by hermit crab

rocks along the coast of Maine where the water is coldest. The larvae of these also settle close at hand and increase the local population without venturing too far afield, though only those that settle in the right niches on the weed, at the right level and locality, are able to survive. Farther down, on the broad blades of the great brown kelp that swirls in the water at low tide from Cape Cod northward on the Atlantic Coast, you find the delicate lacy tracery of *Obelia*, with white branching stalks an inch high rising at short intervals from fine cables glued to the weed. Similar species of *Obelia* decorate the wharf pilings along the Pacific Coast. At times this hydroid gives off innumerable microscopic jellyfish which become mature when about one quarter of an inch across, which scatter the species far and wide. From Cape Cod south to Florida the pilings are likely to be covered with an orange-red hydroid with brown tangled stems (*Pennaria*), though only where the water flows clean and strong. This also produces small jellyfish about one eighth of an inch across, though they are mature at birth and shed their eggs and sperms almost at once and die shortly afterwards. Sometimes the little jellyfish fail to break loose from the parental colony and simply shed their products into the adjacent water. At other times, particularly during summer months at flood tide at dusk, swarms of the small jellyfish can be seen swimming among the piling. If you take a swim in the midst of them you will discover that even so small a kind can sting well enough to make you smart and raise a rash. Vinegar will counteract the poison.

One of the most fascinating hydroids (*Hydractinia*), commonly lives as a pink carpet enveloping mollusk shells inhabited by hermit crabs. Wherever hermit crabs are abundant, from the Carolinas to Labrador, a fair percentage will have this hydroid cloak in addition to their molluscan armor. If you look closely at the living carpet with a pocket lens you will see that the individual polyps are of two kinds, tall whitish ones which are the feeders, and shorter ones with reddish reproductive sacs. Along the Pacific Coast and in places along the Atlantic Coast the same hydroids also live on rocks at low tide. (See question 633.)

810. What kinds of jellyfish are there? Many species of jellyfish inhabit the seas, but only a few are of particular interest to observers on or about the shore. Of the larger kinds the ubiquitous moon jelly (*Aurelia*) and the giant reddish Arctic jellyfish (*Cyanea*), commonly seen with the moon jellies, appear in spring and summer along the Atlantic coast (see questions 818–830); while the lagoon jellyfish (Cassiopeia) lies on its back in sheltered waters among the Florida Keys and the West Indies. (See question 832.) Of the smaller kinds which are legion, generally known as the thimble jellies, many will be seen sporadically in swarms near the surface of smooth waters during spring and summer. Last but not least, although not easy to find, is the peculiar stalked jellyfish (*Haliclystis*) which lives attached by its stalk to the blades of eelgrass and kelp. (See question 831.)

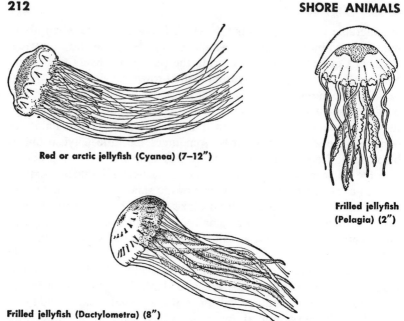

Red or arctic jellyfish (Cyanea) (7–12″)

Frilled jellyfish
(Pelagia) (2″)

Frilled jellyfish (Dactylometra) (8″)

811. How much of a jellyfish is water? More than 95 per cent, but the water is combined with organic substances and mineral salts to form a strong jelly. Some large jellyfish, when stranded on a beach, can be jumped on without being crushed.

812. How does a jellyfish swim? By means of a contractile sheet of muscle near its outer rim. As it gently relaxes, water fills the cavity of the bell; when it contracts, it does so quickly and forcibly and the water is driven out, thereby forcing the animal in the opposite direction. (See question 869.)

813. How large can jellyfish be? Some kinds are fully grown when only one eighth of an inch across. On the other hand some jellyfish grow to be more than eight feet across. (See question 794.)

814. What is the radiating network you see running through the jelly? These are fine branching canals passing from a central cavity or stomach out to a circular canal running around the rim, through which food and oxygen-carrying sea water circulate.

815. How do they feed? The fringe of small tentacles traps small crustacea and other forms, while the four large but highly transparent lobes hanging down from the mouth are covered with minute beating hairs and maintain a steady, food-bearing current of microscopic organisms into the centrally located mouth. (See question 788.)

816. What do jellyfish feed on? Any creature which blunders into the trailing tentacles. Stinging cells on the tentacles paralyze the prey, which is drawn into the mouth at the center.

817. Can all jellyfish sting? All are equipped with stinging cells, but only some are able to penetrate the human skin.

818. What animals feed on jellyfish? Very few, because of their size and general lack of substance. Certain jellyfish, however, feed on other jellyfish (*Cyanea* upon *Aurelia*), while sea turtles to some extent and the giant oceanic sunfish almost exclusively live by nibbling round and round at the edges of large jellyfish.

819. What are the common jellyfish with long tentacles found along the Atlantic coast south of Cape Cod? The largest of the two common forms is the frilled jellyfish (*Dactylometra*) which occurs abundantly in such places as Narragansett Bay and Chesapeake Bay, where the water is pure. It is a beautiful jellyfish with a bell about eight inches in diameter, the margin of which is divided into forty-eight small flaps. Forty thin yellow tentacles stream down from the bell margin about three feet in length, while four long massives lobes or tentacles hang down from the center of the bell, each with frilled ruffles of red or rosy pink. The red color is sometimes lacking altogether. The sting of the tentacles is severe, even the slightest touch producing a burning sensation, while extensive stinging may cause severe illness. (Questions 809, 833.)

The other southern jellyfish (*Pelagia*) is smaller, about two inches across, but is otherwise very similar, with sixteen marginal divisions and about eight marginal tentacles. The four large mouth lobes or tentacles are frilled like those of Dactylometra and are pink. The bell itself is purple rose, shading into blue, and is luminescent at night. This form is one of the most beautiful and graceful of all jellyfish.

**Jellyfish polyp (1/4–1/2″)
constricting off young jellyfish**

820. What is a moon jelly? The moon jelly (*Aurelia*) is one of the commonest and most widespread of all the larger jellyfish. It occurs in the coastal waters of all seas. In colder waters shoals of these jellyfish are sometimes so dense that the water seems to be almost solid with them. It is easily recognized by the shallow saucer-shaped bell, the fringe of numerous small tentacles and the four large horseshoe-shaped white or pink glands near the center.

821. Can moon jellies sting? Well enough to disable the small sea creatures they feed on, but not so that a person can feel it. They can be safely handled.

822. Where do moon jellies come from and where do they go? Each early spring they are produced in vast numbers as small discs about one-eighth of an inch across, budded off from slender polyps attached to the sea floor. They drift with the local currents through spring and summer, growing all the while, until by June or July they often appear as immense shoals. By late summer they have shed their eggs and young and shortly afterwards break up and disappear. The small larvae which develop from the eggs cling to the four lobes below the parental mouth until they are ready to glide away, usually in August or September. Shortly thereafter they settle on the sea floor, become attached and grow into tiny polyps. In this form they live through the winter, and moon jellies as such do not exist. In late winter or early spring the polyps grow longer and each subdivides into thirty or more transverse discs which are cut off in succession from the end, each disc swimming away to become a moon jelly later on.

823. What are the four horseshoe-shaped rings? The reproductive glands, male or female as the case may be. Eggs are fertilized before they escape, by sperm shed by males in the vicinity, and begin their development within the cavity of the horseshoe.

824. Do moon jellies always have four rings? The great majority have four rings only, but a few may have five or six. Those with more than four are those that were the first to be budded from the polyp.

825. How do moon jellies navigate? So far as horizontal movement or travel is concerned, they simply drift wherever currents or tides take them, but they maintain a position close to or not far from the sea surface—depending on how rough it is—by means of eight small gravity-sensitive organs located on the rim and recognizable by notches or interruptions along the fringes of tentacles. Whenever a jellyfish begins to sink toward the sea floor, these tiny organs operate so that the animal is made to swim upward. Otherwise, in overdeep water the jellyfish would sink and get lost in the dark, impoverished depths or arrive at the sea floor in coastal regions and be destroyed.

826. Where is the red or arctic jellyfish found? This is the giant jellyfish of the Atlantic Ocean, which ranges from the Arctic Ocean to the New England coast and is common throughout the North Atlantic. It is found in southern New England about the middle of June, though mature individuals are abundant in the colder water off the coast of Maine in August and September.

827. How large do the red jellyfish grow? The extreme size, encountered only in the open ocean, indicates a truly formidable creature with a large circular or lens-shaped disc as much as eight feet across, with more than 800 tentacles trailing 200 feet. Most individuals, however, are not over three feet across with seventy-five foot tentacles. Specimens one foot across are quite common along the New England coast.

828. What is the stinging capacity of the red jellyfish? The stinging cells are powerful and can penetrate the human skin. They are crowded along the length of the thin tentacles. Swimming in the neighborhood of red jellyfish is dangerous, since the tentacles even of small coastal individuals may extend twenty to thirty feet from the body and are invisible and difficult to avoid. A close encounter with the numerous stinging tentacles has at times proved fatal to swimmers.

829. What are the colored structures of the red jellyfish? The centrally located stomach is rosy pink to brownish purple with a yellowish margin. It has long, thick folded mouth-arms of brownish-purple and hundreds of fine extensible yellowish tentacles. The powerful muscular system which makes possible the rhythmical contractions of the disc is brown to yellow in color.

830. What do red jellyfish feed on? On moon jellies, fish, small crustaceans and virtually everything that drifts in the sea currents. The long, drifting lines of tentacles are effective fishing lines, and the great colored mouth-arms are efficient in handling and digesting any large creatures that happen to get caught.

831. What is the stalked jellyfish? This is a small jellyfish (*Haliclystis*) not more than an inch across, which has the upper central part of its disc or bell drawn out as a stalk which attaches the animal

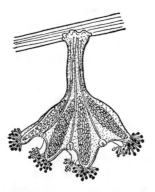

Stalked jellyfish (1″)
attached to blade of eelgrass

to seaweed, either eelgrass or kelp. It cannot swim, and has a central mouth and eight marginal knobby tentacles with which it captures small crustaceans etc. In other words, it lives as though it were a sea anemone rather than a jellyfish.

832. What is peculiar about the lagoon jellyfish? It is one of the laziest jellyfish in the world and spends most of its time lying on its back in shallow lagoons with its mouth side up. Eight complex mouth lobes or tentacles lie within the saucer thus formed, exposed to the light and water. The original mouth is small or absent, and food organisms are taken in through numerous small openings on the lobes. Microscopic algae which live in the lobes make use of the sunlight for their own growth. The relationship between algae and jellyfish is considered to be of mutual benefit. Lagoon jellyfish can swim in the usual manner when the need arises. It grows to a diameter of about six inches.

833. Do Portuguese man-of-wars sting? The man-of-war's sting is notorious. The long tentacles will raise a welt whenever they touch the skin, and the sting is very painful. A badly stung person not only suffers an agonizing pain but also shock and prostration which can be fatal when extreme. Several poisons are thought to be involved, but their nature is not known. A doctor's care is called for, although mild encounters may be treated with vinegar.

It is curious that some fish either are immune to the action of the stinging tentacles or else manage to avoid getting stung even though they swim close under the animal where the tentacles are closest together. The young of cod and butterfish and even the adults of a little blue-backed fish (*Nomeus gronovii*) appear to swim freely among the tentacles, completely safe from attack by others so long as they stay close. It is more likely that these small fish find room enough and are sufficiently aware and agile to avoid getting stung, and that those that do get stung become *Physalia* food like any other victims. (See questions 809, 821, 828.)

834. How did the Portuguese man-of-war get its name? From the early days when all fighting ships were sailing ships. The blue sails of *Physalia,* up to a foot or more in length, often seen in fleets sailing beam to the wind, immediately evoked the impression of small sailing vessels. The name was given to them by English sailors who encountered them first in the seas of Portugal, since north of that they are rarely seen in massed formation.

Portuguese Man of War, showing float
or sail (6") together with long stinging
tentacles and feeding processes

835. Where do the man-of-wars come from? They are oceanic
voyagers belonging to the warmer waters throughout the world, sailing at an angle to the prevailing winds and subject to the vagaries of
the wind. They are also carried along by strong currents and come
north with the Gulf Stream. Strong onshore winds at any time, though
particularly in summer months, drive them from the near-by Gulf
Stream into the bays and sounds and on to the shores of the Atlantic
Coast from Florida to Cape Cod.

836. What is the Portuguese man-of-war? To say what the man-of-
war is can be simple up to a point, but then extremely difficult. It is a
colonial organism made up of numerous semi-individual components
corresponding to those we see in the life history of most hydroids and
jellyfish. The clustered feeding mouths hanging down beneath the
float match the feeding polyps of a hydroid colony (see question
806), though they lack tentacles for capturing their own food. The
long, stinging tentacles also correspond to hydroid polyps, though
they are greatly elongated and lack a mouth. Other units joining the
rest beneath the float are exclusively reproductive, producing eggs
and sperm in season according to sex. All unite at the base with a
common tubular system and nourishment by one is nourishment for
all. The float itself apparently corresponds to a jellyfish type of individual, though very highly modified, and the whole creature can in
fact be compared to a jellyfish dragging its yet attached parental
hydroid behind it without ever breaking loose. The whole is a composite, but a composite so united that all separate individuals are

submerged and a new kind of super-creature emerges as one of the early wonders of the world.

837. What is the float filled with? It is filled with gas which is a little more than 90 per cent nitrogen, a trace of argon and the rest oxygen. In other words it is like air from which half the oxygen has been withdrawn. The total quantity of gas within the float is somewhat variable and appears to be in some degree under the control of the animal, although in the absence of any kind of centralized nervous system it is difficult to imagine what kind of control could exist.

838. How do Portuguese man-of-wars move through the water? Entirely by means of wind. Whereas their less specialized relatives have a very small float which merely keeps one end uppermost within the water and have a number of swimming bells, unmistakably similar to jellyfish, which propel the whole through the water, the man-of-war depends upon its float alone. The float is so large and so filled with gas that it rests almost entirely above the sea surface where it catches the wind. It is shaped like a lateen sail and takes the wind in such a way that the animal sails roughly at a forty-five-degree angle to the wind's direction. Since the ocean currents set generally in the same directions as the prevailing winds, at least at the surface, the man-of-war tends to sweep more or less across the currents. And in some mysteriously remarkable way it seems to trim its sail. The drag of the tentacles in the water keeps it heading more or less into the wind, but the activity of the float itself is astonishing. "Not only can the sail-like crest be held erect or be depressed into a mere ridge, but the whole bladder can alter in shape and twist about. It can roll over from side to side to wet itself on a calm day when there is no spray flying. Also it is a thing of such beauty with its vivid, yet delicate coloring of iridescent blues and pinks, that few can view it without delight."—(D. Wilson.)

839. What are their nearest relatives? The sub-group to which the man-of-war belongs is known as the Siphonophores, a name which refers to the float which all possess. All are composite, colonial organisms of a peculiar kind. The closest relatives of the Portuguese man-of-war are the *Velellas* and *Porpitas. Velella,* which means "little sailor" and often called "by-the-wind-sailor," has a rectangular shape about four inches long with a flat deep blue-green float that has a keel-like crest running diagonally across it. *Porpita* is flat, circular, about one inch across, is a bright blue shading into a beautiful iridescent green in the center, and lacks the sail crest. Sometimes they fleck the ocean for miles with specks of brilliant blue, and both the *Velellas* and *Porpitas* are often carried by wind or current into coastal waters and bays from the Carolinas to the tropics. All the forms just mentioned are the most colorful and specialized of the group, all making use of wind and current for locomotion. The members of the group are rarely seen, are more crystal clear, fragile, and less often

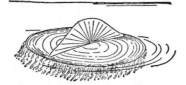

By-the-wind sailor (2″), showing disc and sail

in inshore waters. They may be found, however, among the islands of Maine in early spring, and also among the West Indian Islands and consist of a long chain with a minute float at the top end, then a variable number of rhythmically contracting and expanding swimming bells like those of jellyfish, followed by what is in effect a long fishing line carrying both feeding members and long stinging tentacles. *Physalia, Velella* and *Porpita* differ from this more general kind in being more complexly aggregated and in having enlarged their float and lost their swimming bells, thereby wasting no effort of their own merely to stop from sinking or for moving through the water on their incessant fishing expedition.

840. What are corals? Corals are animals, all more or less closely related to sea anemones. They are of several kinds; the hard or stony corals with a skeleton of lime, the horny corals with a supporting structure of horny material, and the soft corals which have a fleshy consistency and no true skeleton. (See questions 844, 853.)

841. What kind of corals are there? The common stony corals of the Florida Coast and the West Indies are the branching staghorn coral, the massive brain coral with folds like those of the cortex of the human brain, and the star coral, which has a smooth, boulder-like

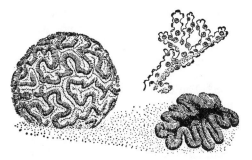

Brain coral (1–2'), fungus
coral (6–10") and branching
coral (6")

surface deeply pitted with star-like depressions, with the finger coral, the ivory bush coral and the fungus or rose coral as runners-up.

The horny corals include the sea fans, sea whips, sea rods, sea pens, and sea pansies as well as the precious coral of the Red Sea and adjacent waters.

Soft coral in the Northern Hemisphere is represented by the finger coral.

842. Where are corals found? The reef-forming or reef-associated corals, such as those seen among the Florida Keys, live only in shallow water where the sea temperature exceeds 70 degrees F. The small stony coral, (*Astrangia*) which forms masses rarely larger than your fist, is found however as far north as Cape Cod; while the single-cupped solitary corals live even further north and are not uncommon in rock pools of central California. (See question 847.)

843. What is the living part of a coral? The thin film of soft substance covering the hard skeleton. To see it properly, coral should be

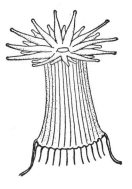

Soft, living, part and hard skeleton of a coral polyp

observed at night when its tissues are usually expanded and each coral unit is then seen to be a delicate translucent column surmounted by a ring of tentacles. When exposed to light or otherwise disturbed, the living tissues draw down so as to be hardly detectable.

844. What is the coral skeleton? The coral individual is like a sea anemone in which hard lime has been laid down in the circular wall and in the radially arranged membranes of the interior. When the living substance dies or contracts, the white or creamy limestone skeleton is all that remains to be seen. "These dried and bleached skeletons are beautiful and are used as ornaments. But they give about as much impression of the exquisite beauty of living, expanded corals as one would get of the beauty of a woman from her whitened bones." —(R. Buchsbaum.)

845. How do corals live? Primarily they live like small sea anemones, feeding by directing food organisms to their mouths by means of the tentacles and outer surface generally. In the case of the northern stone coral of the Atlantic Coast (*Astrangia*) which has been studied more carefully than most because of its proximity to the great Marine Biological Laboratory at Woods Hole, the coral feeds in the natural state on diatoms, small larvae of crustacea, etc., or anything in fact which is small enough to be drawn into the mouth with the slime to which it becomes attached or which can be stung and manipulated by the tentacles. Feeding occurs mainly by night in all corals. Besides being sensitive to light, the coral individuals are highly sensitive to the presence of living material that might serve as food. Tissue juice from a mollusk, even when diluted more than one thousand times with sea water, will cause a coral polyp to expand. Corals as a whole, however, have a trick up their sleeves which makes life easier for them. Their tissues are for the most part loaded with living microscopic plant cells or unicellular algae which dovetail their metabolism with that of the polyps. The polyps use up oxygen and give off a variety of waste products, including carbon dioxide. The plant cells use up the waste products and give off oxygen during their photosynthesis, so that the two members of the association, so to speak, take in each other's washing. One result is that corals can live in much more crowded quarters than would be otherwise possible, and their association with their numerous but microscopic plant guests is an important factor in making coral reefs, as distinct from isolated corals, possible.

846. Are corals always in the form of massive or branching colonies? Most corals consist of numerous individuals joined together, both by their living tissue and by their skeletons, to constitute large colonial structures, but a few remain as single solitary individuals throughout life. With reference to the rocky Pacific Coast, Ricketts and Calvin write: "The vividly orange-red solitary coral, (*Balanophyllia elegans*), is abundant, but must be well sheltered from desiccation and direct sunlight. In late winter (February and March) at Pacific Grove, the transparent and beautiful cadmium-yellow or orange polyps may be seen extending out of the stony, cup-like base until the base comprises only a third of the animal's bulk. Large individuals may be one-half inch in diameter expanded. The range is from Puget Sound to Monterey Bay." Among the Florida Keys the specimens of the fungus or rose coral, as large as your hand, are also single individuals, although in its youth each individual has given rise to a number of individuals through successive separations of the head down the stalk, each product subsequently growing to maturity without much further confusion as to who is who except that all solitary corals, throughout their life, produce free buds from almost any part of their surface. (See question 806.)

847. How do corals grow and multiply? Corals are truly animals and, like most others, are sexual creatures of separate sex in 'spite of appearances. Eggs are shed into the surrounding water in season and develop into microscopic gliding larvae which settle on the sea floor after a week or so of drifting. By this means corals slowly become dispersed throughout the range of temperature, light and depth which they can tolerate.

They also multiply by budding new individuals somewhat in the manner of certain sea anemones (see question 797). In the case of solitary corals such as the orange cup coral (*Balanophyllia*) and fungus coral (*Fungia*), the buds separate from the parent and become independent individuals. In the rest, which are the majority, the buds remain attached to the parent in the manner of Siamese twins, and the process results not in any increase in number of corals but in the growth of the original as a massive or branching structure bearing numerous polyps all in some degree of communication with the common stock.

848. Can corals live out of water? Only for short periods of time. They are rarely found exposed except at very low tides. If taken out of the water, the delicate living material very quickly turns into a slimy film.

849. Are corals harmful? Some corals sting with unpleasant although not dangerous effects (see question 809). Most corals, if carelessly handled, stepped on or otherwise encountered, easily lacerate the skin. For reasons somewhat obscure, the cuts usually heal with much difficulty.

850. What use is made of corals? The hard, lustrous, reddish precious coral of the Red Sea and elsewhere is used for jewelry; in Bermuda coral limestone is sawed into building blocks.

851. What is precious coral? The red skeleton of one of the horny corals commonly found in the Mediterranean and elsewhere, used for the making of jewelry.

Precious coral (1/2″)

852. What is a coral reef? Coral reefs are constructed mainly of the dead skeletons of hard corals, with living corals growing on the reef near the surface. As they die, their skeletons are added to the preexisting mass and so the reef grows. Most of the active growth of the reef is on the outer or exposed side where the slope is steep and the surf is breaking, for this is the side where food and oxygen are most abundant. Dead coral, however, is continually being tossed over the top of the living reef to accumulate behind, so that with time a reef widens at the top and the living front shifts gradually seaward. On the outer side where the corals are exposed to strong wave action, most of the corals are of the massive or short-branched kinds, while the taller slender-branched kinds are usually found in the sheltered waters behind. Many other creatures contribute to a reef, however, including tube-building worms, stony algae, and the host of other forms which find shelter or support on the reef, such as the sea fans, sea whips and many others.

853. What is a sea fan? Sea fans are horny corals and like all such consist of numerous polyps, each with a mouth and eight tentacles.

Sea fan (1–2′)

Instead of forming a heavy skeleton of lime like that of the hard or stony corals the living tissue is supported by a branching core of tough horny material which gives the colony both stiffness and resilience.

854. Of what advantage is the shape of the sea fan? By presenting the broad face of the fan to the direction of the prevailing current, the colony sifts a maximum amount of water for the small food organisms it contains.

855. What happens if you take a fresh specimen of sea fan home? You will have one of the most disagreeable smells in creation, and one that lasts for weeks. (See question 54.)

856. What are sea whips? Sea whips are very similar to sea fans, differing mainly in the absence of cross struts which link the main branches together.

857. What is a sea finger? It is a form of soft coral, related to sea fans but lacking an internal skeleton. Soft corals are spectacular and common inhabitants of coral reefs. Along the Atlantic Coast the orange sea finger may be seen inhabiting the under surface of rocky ledges exposed only at extremely low tides.

858. What is a sea pen? "Sea Pens, which belong to the same class as the anemones, have a bulb or stalk that fastens them in the mud of the ocean bottom and a stalk that extends above the mud. This upper portion of the stalk is covered by a colony of polyps that feed the organism. Some sea-pen colonies are short and heavy and shaped like an ostrich plume. If feeding conditions are unsatisfactory, they can remove their peduncle from the mud and move to another location. Other sea pens are long and slender, reaching a length of over six feet."—(G. E. and N. MacGinitie.)

859. Where can sea pens be seen? Most sea pens live anchored to the bottom in comparatively deep water and are generally luminescent. Along the coast of North America they can be seen in shallow, even tidal waters in flats of sandy mud in Newport Bay, California. These are individuals of the species *Stylatula elongata*.

"The collector who appears on the scene between half tide and low tide will be willing to swear before all the courts in the land that he is digging out a thoroughly buried animal. But let him row over the spot at flood tide, and he will see in the shallow water beneath his boat a pleasant meadow of waving green pens, like a field of young wheat. The explanation comes when he reaches down with an oar in an attempt to unearth some of the 'plants.' They snap down into the ground instantly, leaving nothing of themselves visible but the short spiky lips of their stalks. Like the anemones (*Harenactis*), they have a bulbous anchor that is permanently buried deep in the bottom. When molested, or when the tide leaves them exposed, they retract the polyps that cover the stalk and pull themselves completely beneath the surface. When cool water again flows over the region they expand slowly until each pennatulid looks like a narrow green feather waving

Sea pen (18")

in the current . . . There must surely be more than a hundred in a square yard."—(E. K. Ricketts and J. Calvin.)

860. What is a sea pansy? The sea pansy (*Renilla*) is closely related to the sea pens, although it lacks the horny supporting rod, and together they form a particular group of the so-called soft corals.

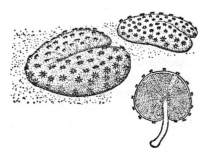

Sea pansy (1–2")

861. How can sea pansies best be seen? "When seen in their natural habitat, sea pansies seem misnamed, for they lie with their heart-shaped discs almost covered with sand, their stalks buried. A specimen should be transferred to a porcelain or glass tray containing sea water and left undisturbed for a time. The disc will then enlarge to three times its size when taken from the sand, and scores of tiny hand-shaped polyps, beautifully transparent, will expand over the purple disc. Not the least of the sea pansy's charms is the blue phosphorescent light which it will almost invariably exhibit if mildly stimulated with a blunt instrument after being kept in the dark for an hour or so. In Newport Bay, sea pansies occur by the hundred, in El Estero de Punta Bunda by the thousand. . . . Southern California is the northernmost part of the range."—(E. K. Ricketts and J. Calvin.) They are also common in shallow water among the sea islands of Georgia and the Carolinas.

862. How do sea pansies feed and change their shape? "The various individuals or polyps all have the shape of little tubes embedded in the fleshy substance of the colony. But some of the tubes bear tentacles and look like small sea anemones; these capture food for the colony, and in proper season form reproductive cells. Other tubes lack tentacles; these are the engineers of the colony, attending to the functions of water-intake and control. A hydraulic system of changing water pressure controls the movements of the colony; as the stem is made turgid it may be thrust down into the sand, drawing the main body after it."—(Rachel Carson.) (See question 168.)

CTENOPHORES

863. Where are comb jellies found? Through the oceans of the world and in numbers beyond comprehension. Being feeble swimmers they are carried helplessly by currents and tides, and swarms of them may at any time occupy a particular bay, inlet or sound for a few days only to be gone a little later. They are frequently cast up as glutinous blobs on the sand at the water's edge. They are best seen from a skiff when the water is as smooth as glass, for then they come closest to the surface.

864. What other names are given to comb jellies? The scientific name is ctenophore, meaning comb-bearer. Certain small forms are called sea gooseberries because of their general appearance. Others are known as sea walnuts. A large, belt-like tropical form is called the Venus girdle.

865. What kinds of comb jellies are there? Three kinds are commonly seen: the small sea gooseberries, (*Pleurobrachia*) with a pair of extensible tentacles, less than one inch, often occur in great swarms along both Atlantic and Pacific Coasts; the somewhat larger sea walnuts (*Mnemiopsis*) up to four inches long, with flap-like mouth lobes common south of Cape Cod, particularly in Long Island Sound, almost invisible by day but marvellously luminescent at night; and the dirigible-like species of *Beroë* which grow to a length of six inches, common on both coasts.

866. What are the eight ridges along the sides of a comb jelly? They are rows of ciliated combs which serve for locomotion.

867. Are comb jellies related to jellyfish? No. In spite of their gelatinous nature comb jellies are set apart as an independent group of animals with no known relatives.

868. Can comb jellies sting? No.

869. How do comb jellies swim? By means of eight rows of ciliary combs which radiate over the surface of the animal from the upper pole to the lower pole, like the lines of latitude on a globe. The eight rows beat in unison and the animal is slowly propelled through the water. (See question 812.)

870. Do comb jellies produce light? "The rapidly beating combs refract light and produce a constant play of changing colors. Comb jellies are noted for the beauty of their daytime iridescence, but this is certainly matched at night by those comb jellies that are luminescent. When the animals are disturbed as they move slowly through the dark water, they flash along the eight rows of combs."—(R. Buchsbaum.)

871. How do comb jellies feed? By swimming slowly with open mouth forward, some forms trailing a pair of sticky tentacles, others with large lobes about the mouth or just with a very wide mouth.

872. What do comb jellies feed on? The more typical comb jellies feed on tiny crustaceans, floating fish eggs and the larvae of mollusks, crabs and echinoderms. As many as 126 oyster larvae have been found in the stomach of a single specimen. The large mouthed, dirigible-shaped species of *Beroë,* however, feed upon other comb jellies and some small jellyfish among which it swims.

873. How do they reproduce? Comb jellies are all hermaphrodites, combining the two sexes in the same individual. Eggs and sperm are shed into the water and the eggs develop directly into small comb jellies. Unlike jellyfish, they have no stage of their life history associated with the sea floor but are oceanic throughout life.

SPONGES

874. Are sponges animal or vegetable? They are animal, although: "It was not until a hundred years ago that the last sceptics were finally convinced of the true animal nature of sponges. The question which they had been asking and which previously had not been satisfactorily answered was: 'Since sponges do not move about and apparently do not respond rapidly to conditions about them, how can they capture food?' This question is readily answered by adding a suspension of fine particles to the water near a sponge, thus disclosing a great deal

Breadcrumb sponge (2–15″)

of unsuspected activity. A steady jet of water is seen to issue from one or more large holes at the top of the animal. Closer inspection reveals that water is at all times entering through microscopic pores that riddle the entire surface. The sponge lives like an animated filter, straining out the minute organisms contained in the stream of water that passes constantly through the body!"—(R. Buchsbaum.) Hence the name Porifera or "pore-animal" which is the scientific name given to the sponges as a whole.

875. Where are sponges found? On the sea floor from the intertidal region to the greatest depth. On rocky shores at low-tide level and below, almost every rock will bear encrusting sponges on its undersurface and so will the underwater margins of low-level rock pools, and rock walls where the sea swell keeps them perpetually wet. Wharf piles and all submerged structures below low-tide level are usually

Finger sponge (10″)

similarly encrusted. Larger, tubular or finger-like branching sponges live in quiet water attached to piling, etc. More massive though squatter types are often found on mud flats. The largest sponges such as the loggerhead sponges, live anchored in sand in shallow water among the Keys and elsewhere in the south.

876. What is a sponge individual? This is a question which has long been argued. In the cases of sponges with a very definite form, like the small purse sponges found hanging beneath rock surfaces, or the large barrel-shaped loggerhead sponge, the whole sponge is clearly an individual. In the case of an encrusting sponge like the bread-crumb sponge, which often covers yards of rock, the answer is not obvious. Some biologists consider the whole mass to be one individual, others say each oscular opening represents the individual and the whole is a colony. You can take your choice.

877. Do sponges have skeletons? Yes. As in the case of most bulky creatures their tissues require some support. In sponges such support is in the form of a fine scaffolding made up either of numerous inter-

Spicules (1/500″) which form sponge skeletons

locking spicules, or crystalline rods, which are either calcareous or siliceous, or else the crystalline spicules are absent and their place is taken by a tough fabric of horny fibers. Those with siliceous skeletons are generally known as glass sponges and are mostly inhabitants of deep, still waters. Their skeletons may exhibit remarkable beauty. Sponges used commercially are and always have been those with a horny skeleton. (See question 879.)

878. What are the holes that you see on the upper part of a sponge? The holes that are large enough to be seen with the naked eye are called oscula and are apertures through which water is constantly escaping after having passed through the sponge wall.

879. What is a bath sponge? The old-fashioned bath sponge is the skeleton of a horny sponge. Suitable sponges have been collected by divers in the Mediterranean since the greatest antiquity, and off the coast of Florida and the Bahamas since 1849. The Florida sponges, coarser than the Mediterranean, are of several kinds: yellow, velvet, glass and glove sponges. The living sponges are usually black and slimy, and must be laid out on deck or land and stamped upon or beaten; then they are strung up and allowed to macerate. Then they are beaten again, washed in sea water and hung up to dry. Finally they are spread out to bleach in the sun. (See question 877.)

880. Why are sponges so differently shaped? The natural shape of the sponge seems to be that of a vase, but such a shape is only possible in very sheltered water, for otherwise the sponge is too easily dislodged. It is the common shape of deep-water sponges and of several very small-sized species of shallow water which can find suitably sheltered waters. Wherever there is wave action, a flat encrusting shape which permits a maximum area of attachment and offers little or nothing to oppose the action of waves and currents is obviously advantageous, as along most exposed or partly exposed rocky shores. Consequently most sponges of such regions are flat, encrusting species well adapted to their mode of life. On pilings and rocks in sheltered regions, below low tide level, a compromise is possible, and branching, finger-like species are commonly found.

881. What are the colors of sponges? Deep-water sponges have little color. Sponges living in the open in shallow tropical seas such as the Florida Keys and the West Indian Islands are usually almost black, the pigment serving as protection against the intense light. Under rocks and reefs and in sheltered rock pools, sponges may be red, green, yellow, or even blue, though the color seems to signify little.

882. How do sponges feed? By filtering water. (See question 874.)

883. What does a sponge feed on? On microscopic organisms drawn in with the water through the millions of minute pores which cover the surface of the sponge.

884. How do sponges reproduce? "Practically all sponges are hermaphroditic, but the eggs and sperm may develop at different times. Probably all marine sponges reproduce by means of eggs which are fertilized within the body of the sponge and develop into tiny larvae that go out with the excurrent water. After a period of swimming about in the ocean they become fixed and metamorphose and grow into the parent type.

Some sponges reproduce by budding off portions of the body. Any piece of sponge is capable of developing into a complete sponge under favorable conditions, but the process is extremely slow. In the majority of sponges the bud does not separate from the parent body, and budding serves only to enlarge the colony."—(G. E. and N. MacGinitie.)

885. How long do sponges live? Some species live only for a year or even less, but some of the large slow-growing leathery sponges that live in quiet waters probably live to be twenty to fifty years old.

886. Can sponges live out of water? Not at all. Their whole existence depends on a continual stream of clean aerated water passing through their tissues. (See question 874.)

887. Do sponges live only in the sea? The very great majority of sponges are marine, though one family of horny sponges is exclusively fresh-water.

888. What enemies do sponges have? Sponges have few enemies, although certain sea slugs feed on them. The slugs usually are of much the same color as the species of sponge they feed upon. The spicules of calcareous and glass sponges render them inedible to most animals, while the disagreeable odor, which may be extremely offensive to humans, may be protective generally. (See question 364.)

889. Have they any economic use? Certain kinds of horny sponges are still used as bath sponges, although synthetic substitutes have now for the most part taken their place. "Sponges, or rather the skeletons of sponges, were commonly used by the ancient Greeks for bathing, for scrubbing tables and floors, and for padding helmets and leg armor. The Romans fashioned them into paint brushes, tied them to the ends of wooden poles for use as mops, and made them serve on occasion as substitutes for drinking cups. Today, sponges have an even wider variety of uses, and 'sponge-fishing' is an industry which every year produces over one thousand tons of sponges. Bath sponges grow only in warm shallow seas."—(R. Buchsbaum.)

890. Can sponges be cultivated? Yes. In the Bahamas, especially before sponge disease more or less ruined the industry some time ago, live sponges were cut up and each piece fastened securely to a small rock, and replaced in shallow water. Within two or three years a well-shaped, fair-sized sponge would reform from each piece. Some crabs, in fact, have been cultivating sponges for time out of mind, placing pieces of sponge on their backs and legs where they continue to grow until the whole crab is covered with the sponge.

891. Where is the Florida sponge fishery centered? At Tarpon Springs on the Gulf Coast. It is carried on by Greek fishermen or their descendants, from ships of typical Greek design and colors. The sponges are gathered by divers from offshore banks in fairly deep water.

892. Why does a sponge so often contain so many small creatures? Because the cavities within a sponge offer ideal shelter—narrow, dark and yet always well supplied with a stream of clean water which also carries microscopic food organisms. Small crustaceans and worms are the commonest inhabitants. (See question 893.) Brittle stars wander in and out in search of them. (See question 199.)

893. What shallow-water or tidal sponges are of particular interest? 1. The loggerhead sponge (*Speciospongia*) of the Florida Keys, notable for its black, barrel-like bulk and for the host of its inhabitants. Large individuals harbor thousands of small pistol shrimps, which are lobster-like in shape and have one large snapping claw, together with true crabs and numerous worms. It constitutes a community by itself, comparable to a tide pool. In one case over 12,000 pistol shrimps were found inhabiting the large maze of sponge tissue.

2. The sulphur-yellow boring sponge (*Clione*), found on both Atlantic and Pacific Coasts, penetrates shells and limestone rocks like a fungus, finally protruding from the surface to form thick, irregular crusts several inches in diameter. Boring sponges are the chief agents in the disintegration of shells that accumulate on the sea floor, responsible for much of the conversion to shell gravel and finally to shell sand.

3. The mud-flat sponge (*Tetilla*) which lives on the soft surface of mud flats in estuaries such as Newport Bay along the coast of southern California, a glass sponge which anchors itself in the mud by means of root-like tufts composed of spicules; and the gray sponge (*Suberites*) about as large as your hand, which occurs in muddy pools at low tide along the Carolina coast.

4. The green or yellow-green bread-crumb sponge (*Halichondria*), so named because of its crumby nature, which lines the sides of cool rock pools and overhung rocks near low tide and forms a spongy carpet punctuated every inch or so by oscular openings for the escaping water, a happy hunting ground for brittle stars.

ASSORTMENT

894. What are sea mats? They are individually minute animals which are united to form encrusting or bushy colonies—the group being known as Bryozoa. The bushy types in some cases look very much like hydroids (see question 804), but are in no way related to them.

895. Where are sea mats found? The encrusting kinds (especially the species of *Membranipora*) form encrustations on the blades of kelp and shore weeds and under rocks along both Atlantic and Pacific Coasts and the shores of Northern Europe. The colonies look like frost tracery or very fine close lacework on the darker weed. A hand lens brings out the beauty of design in a striking manner, and the tracery is seen to consist of a flat, branching system incorporating minute holes. Each hole opens into a small chamber occupied by an individual.

The bushy forms are equally abundant on every coast, attached to the sides and undersurfaces of rocks etc. at low-tide level (principally species of *Bugula*), forming brownish to pale yellow spiral tufts an inch or two long. Both the bushy and encrusting kinds commonly attach to wharf piles and to the bottom of ships.

896. How do sea mats feed? Each individual, which is attached by muscles to the wall of its tiny shell, extends a horseshoe-shaped circle of tentacles through the opening and feeds upon such small organisms as come within reach.

897. Where are lamp shells found? On the Atlantic Coast only one species (*Terebratulina*) can be found in shallow water near the low tide level, and only along the northern part of New England. The same genus, if not the same species, however, is known as one of the commonest fossils in rocks four hundred million years old—a conservative creature, to say the least. On the Pacific Coast a somewhat similar form (*Terebratalia*) is a feature of the rocky, lower intertidal fauna of British Columbia extending to Puget Sound. A long stalked kind (*Glottidia*), however, lives locally in large numbers in muddy sand in southern California, for instance in Newport Bay.

898. How can you distinguish a lamp shell from a bivalve mollusk? Where the two shells come together at their narrow end, one extends beyond the other and has a small hole in its end where the muscular stalk passes through which attaches the animal to rock or some other object on the sea floor. There is no hinge.

899. How do lamp shells live? The body, protected by its pair of shells, is attached by a muscular stalk, short in the rock-attached kinds, long in the mud and sand dwellers and in some cases virtually absent. The stalk can lengthen or shorten and the body may twist upon it in one direction or another. The animals feed by drawing in a current of water through a narrow slit formed between the two shells, filtering out the contained minute organisms.

900. What are lamp shells? Lamp shells constitute an exclusively marine group of animals known as the Brachiopods, which in spite of superficial resemblance are in no way related to mollusks. They do possess a bivalve shell, but there the similarity more or less ends.

Lamp shells (1/2–1″)

Unlike the mollusks they possess a complex internal calcareous skeleton in addition to the two shells, this skeleton serving as a support for a coiled double horseshoe-shaped ridge of tentacles which comprise the feeding organ.

901. What is a phoronid worm? A peculiar, enigmatic worm-shaped creature that lives in tubes of its own making, and seemingly has no close kin in the whole animal kingdom. The slender body has a horseshoe-shaped circlet of tentacles by which it feeds when protruded from the tube. One species (*Phoronis architecta*), about five inches long but no thicker than a pencil lead, is common in the sand flats of North Carolina, in isolated tubes covered with sand grains, but the casual observer is more likely to encounter this creature along the more sheltered parts of the Pacific Coast, particularly on sandy mud flats of the low-tide level, if you are inclined to wander in such sticky places. Both orange and green species may be found, the orange form (*Phoronopsis californica*) lives singly here and there in the estuaries in sand-impregnated tubes from eight to eighteen inches long. "The entire body is orange, with bright-orange tentacles that are often extruded from the burrow and left lying on the sand" . . . the green form (*Phoronopsis viridis*) "lives in sandy tubes from four to eight inches long and about one eighth inch in diameter. It is an estuarine dweller that is often found abundantly in bottoms with a large percentage of sand. At Elkhorn Slough, Calif., we have seen colonies of this species almost an acre in extent. The animals are so plentiful that when their tentacles are extended the bottom of the Slough appears green. When they are uncovered at low tide they draw downward into their tubes, leaving tiny holes in the sand. In one colony we counted

281 individuals in an area four inches square. The tubes help prevent the washing away of the mud."—(G. E. and N. MacGinitie.)

902. Where are acorn worms to be found? Buried in sand at and below the low tide level. A small kind (*Ptychoderma bahamensis*) about three inches long is common in places throughout the West Indies and the Lower Florida Keys. A slender, six-inch form (*Dolichoglossus*) burrows in sand or sandy mud from Massachusetts Bay to North Carolina; it is a colorful type, with a pink proboscis and orange-red collar, and a mottled orange-yellow body. A very closely related form (*Dolichoglossus pusillus*), similarly colored although growing to two feet in length, used to be very common in the bays of southern California, but pollution and harbor improvements have made them scarce. A giant form (*Balanoglossus brooksii*) lives in the extensive exposed low-tide sand flats of the North Carolina coast.

903. How can giant acorn worms be found? The opening of the burrow is readily seen on the sand flats, and only the burrow of the lugworm is likely to be mistaken for it, since these alone appear to make castings at the surface (see question 787). The acorn worm however can be identified by putting your finger in it and then smelling your finger. If it has an unmistakable smell of iodoform or iodine like that of a hospital, an acorn worm lives down below. If you really want the animal, prepare to dig hard and fast with a spade. Then with a little luck you should manage to get the front end of it. The rest may come along piece by piece.

904. How do acorn worms feed? They swallow sand with its contained organisms and other nutritive matter, digesting the latter and passing the indigestible sand to the surface as a cast at the back opening of the burrow.

905. What are acorn worms? They are worms only by virtue of their shape, which is elongated and worm-like, for they are in no way related to any of the groups of true worms. Such relatives as they have are of doubtful connection, and they are thought to be connected with echinoderms on the one hand and with vertebrate animals on the other. If this last be true, then they are our own most distant kin, all appearances to the contrary notwithstanding. The body of an acorn

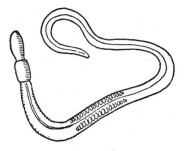

Acorn worm (3–12″)

worm is divided into three parts, a proboscis, a collar, and a trunk region. The proboscis usually has the shape of a high dome and is set at its base within the collar, very much like an acorn in its cup—hence the common name. Acorn worms have genuine gill slits along the sides of the trunk, which is a feature shared with all fish and their known relatives but with no others.

906. What are lancelets? They are famous throughout the zoological world because they clearly possess the forerunner of a backbone together with the muscles and gills of the fish-like animals, but do not have a brain or recognizable sense organs—only a spinal cord without a head! They live in sand, moving through wet-packed sand almost as fast as fish move through water, by virtue of the powerful muscles and the sharp-pointed ends of the body. Usually, however, they lie imbedded in the sand with their mouths at the surface, taking in water and its contained nutritive particles and continually filtering the water out again through the gill slots at the sides of the body, in much the same manner as a fish feeding by means of gill rakers.

907. Where can lancelets be found? The lancelet (*Amphioxus*), which is so abundant along parts of the coast of China as to support a commercial fishery, lives in shallow bays with clean sandy bottoms from the Chesapeake to Florida, although it is rarely taken. It is apparently more abundant in southern and northern Lower California, from Newport Bay to the Gulf of California. Violent stamping on the packed sand at low water in the right area is said to cause some of the animals to pop out, but they dive back in so fast it is hard to believe one's eyes. Likewise, turning up spadefuls of sand will expose them if they are present, but you need to dump the sand in a sieve if you hope to collect any.

908. Do lancelets have eyes? They have no eyes, but the front end of the nerve cord is light-sensitive.

909. What is a sea squirt? Sea squirts are an exclusively marine group of animals of a most ambiguous nature. Either as single solitary forms or as members of a colony, each individual has two siphons or openings for the intake and output of water. When disturbed, the individual contracts and water squirts out, hence the common name.

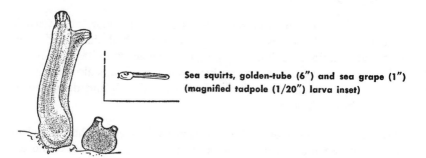

Sea squirts, golden-tube (6″) and sea grape (1″)
(magnified tadpole (1/20″) larva inset)

As a group they are known as the ascidians, which is a sub-group of the tunicates or tunic-covered forms. While the adults may look more like vegetation, anything from fungus to a potato, they produce larvae which are unmistakably related to fish and all other backboned creatures, including ourselves. Their scientific interest is accordingly very great.

910. What are the principal kinds of sea squirts? As far as superficial appearances go, there are three main kinds; the comparatively large, single or solitary individuals; the social types where numerous more or less separate individuals are united at their bases to form a loose colony (not to be confused with mere accidental clusters); and compound forms which consist of a solid mass or thick layer of tough jelly in which large numbers of very small individuals are embedded. This used to be the basis for classifying ascidians but is now discarded in favor of the actual nature of the individual irrespective of whether or not it is united with others.

911. Where do sea squirts live? As a group they range from the intertidal region down to the greatest depths of the sea. Within the intertidal region, however, they live in very different places according to their own nature. The flat, encrusting, compound sea squirts live attached by their broad base to the underside of rocks, the sides of shaded wharf piles, the undersurface of ships, and to the submerged roots of mangrove trees; some are commonly found attached to the blades of kelp and other seaweeds, while others again, though of more massive type, live embedded in firm sand below the low tide level. The so-called social kinds generally require a little more vertical space and are more likely to be found on the sides of rocks and pilings, particularly on submerged wrecks, whereas the large solitary sea squirts live attached to the sides and often the upper surfaces of rocks where they have freedom to grow, though as a rule where the light is not excessive, generally below low water in sheltered regions. Between the tides, at the lower levels, such kinds may however be found attached to the underside of rocks where there is space enough for the animal and enough of a water current to maintain it. Coast lines everywhere have a diverse sea-squirt fauna, although each region has a fauna of its own.

912. What is the tough outer covering of a sea squirt? It is a protective layer called the tunic or test and in the larger forms may be quite thick. The true skin lies beneath it, although the tunic must be considered a living tissue. In some cases it is like a stiff jelly, in others almost like leather, with every consistency in between. Chemically it is an unusual material for an animal to produce, for the principal substance is pure cellulose, typically a product of the plant kingdom.

913. How do sea squirts feed? Sea squirts all feed in one and the same way. Water with the food particles it contains, dead or alive, is drawn through the mouth opening (otherwise known as the inhalent siphon) into a relatively large gill basket or throat. Here the particles of food are trapped by mucous which continually moves across the walls of the basket and then passes back to the intestine, while the water passes through the very numerous tiny gill slits in the walls and finally escapes through the other opening (or exhalent siphon). The escaping water current also carries away the feces passed from the hind end of the intestine, together with eggs and sperm during spawning periods. (See question 375.)

914. Where can large individual sea squirts be easily seen in abundance? Drab-colored sea squirts somewhat like small bumpy potatoes one to three inches across (species of *Styela*) cluster in great numbers on the roots of mangrove trees and on pilings along both coasts of Florida and territory immediately to the north. A larger kind, five to eight inches high (also a *Styela*), is an abundant piling form along most of the Pacific Coast.

A jet-black elongated kind (*Ascidia*) attached by its side rather than by its base, is a very obvious form to be seen on the sides of rocks and submerged stone walls in southern Florida and through the West Indies. Individuals, however, do not cluster. The tunic is not only black but smooth and shiny. The black pigment is a protection against the intense tropical light. Related species of *Ascidia,* pale yellowish-green and translucent, can be found attached to the undersides of small boulders at low tide.

A third kind (*Ciona*) that appears to have no common name but might well be called the golden-tube sea squirt, is a cylindrical form four to ten inches long and an inch or so wide. It typically attaches to wharf piles and floats and to ship bottoms and is now a cosmopolitan sea squirt found in the habors of temperate seas throughout the world. It is or has been locally common along the Atlantic Coast and is an abundant form on the Pacific Coast. The tunic is yellowish-green and translucent so that much of the inner organs can be seen, while the long siphons are green and glossy.

915. What are sea grapes? They are certain kinds of round or ovoid sea squirts (species of *Molgula*) from one-half to an inch or two across, which are firm and grape-like in appearance and touch when disturbed and contracted. When expanded, the two siphons are prominent and easily seen. Sea grapes live attached to the sides of wharves, pilings and submerged woody structures of many kinds along most of the Atlantic seaboard.

916. What is the sea peach? A large, colorful, spectacular sea squirt (*Halocynthia pyriformis*) of much the same size and appearance and velvety texture as a ripe peach. It even has a distinctive pleasant, rather sweet smell. Sea peaches are common in shallow water attached to the sides of rocks from eastern Maine more or less all the way to the North Pole. In the colder waters north of Bar Harbor the sea peach can be found at extreme low tide on rocks at the water's edge. A closely related species of similar size (*Halocynthia johnsoni*)— which, however, has a bumpy tunic covered with all sorts of foreign material—forms almost solid coatings on wharf piles in San Diego Bay and elsewhere along the Pacific Coast.

917. What is the sea potato? It is a large sea squirt (*Boltenia*) belonging to the same family as the sea peach and occupying much the same territory, although it is likely to be seen only by examining permanently submerged wharf piles along the Canadian Atlantic Coast, or when dredged up with the large northern scallop by fishermen. The animal looks like a reddish-brown potato with a slender stalk a foot or more long which is used either to raise the potato-shaped body high above the sea floor or to suspend it from some other surface such as piling.

918. What is the golden star sea squirt? The golden stars are flat encrusting compound sea squirts (*Botryllus*) which cover the broad surfaces of many kinds of seaweed, wharf piles, bottoms of ships and similar places. The colonies are often extensive, as broad as the palm of your hand but only one-eighth of an inch thick, and several colonies of various colors may adjoin one another edge to edge. The individuals within a colony are typically arranged in star-shaped groups, usually of striking appearance. The star individuals generally are picked out in white and yellow and show up against the purples, greens and blue of the test substance.

A related form (*Botrylloides*) in which the individuals are arranged in long meandering branching double rows instead of stars, is equally prominent on the Pacific Coast, and the reds, purples and yellows are striking features on the pilings in San Diego Bay.

919. What is sea pork? Sea pork is commonly found in chunks about the size of a fist, often tossed up along the shore after storms. They are actually rather massive compound sea squirts (*Amaroucium*) that live attached to the surface of stones in shallow water below the level of low tide, ranging from Cape Cod to North Carolina. The colonies have a tough, translucent test which suggests salt pork in color and consistency.

920. How are the colonies of sea squirts formed? Many different kinds of sea squirts produce buds from the sides of the body or else subdivide the body itself like a string of beads, each unit becoming a new individual. Individuals thus produced remain within the original tunic substance, so that instead of this method of reproduction resulting in numerous independent individuals, the result is a colonial mass with many small ascidians more or less joined in communal life. The consequences are clear, for whereas a solitary sea squirt is a somewhat ovoid or cylindrical creature which is increasingly exposed to the action of waves and currents as it grows, a colonial or compound sea squirt grows by increasing the number of its constituents and therefore mainly in a horizontal direction—i.e., it extends the area of attachment without expanding outward into the water. (See question 806.)

921. Where are social sea squirts? Flower-like clusters of white and crystal forms (*Clavelina*) or orange (*Ecteinascidia*) may be found attached to the sides of rocks, submerged piling, corals and mangrove roots, etc., among the Florida Keys. The former also occurs on the California coast as great colonies in the lower rocky tide pools around Monterey and elsewhere. The individual members in each kind are about one inch long.

922. Do sea squirts have sexes? All sea squirts, large and small, solitary or colonial, are simultaneously male and female, i.e., are true functional hermaphrodites. Cross-fertilization of eggs is the general rule, however, in spite of the possibility of self-fertilization.

923. How do sea squirts reproduce? All produce eggs and sperm, and the fertilized eggs develop into small free-swimming tadpole-like larvae with a swimming tail, nerve cord and an eye and a balancing organ. After a few hours of swimming toward the sea surface, the tadpoles sink to the sea floor and attach themselves by means of cement organs on their front ends, resorb their tails much as in the case of frog tadpoles, and grow into typical adult sea squirts.

924. Are sea squirts of any use? Not directly. Indirectly they appear to be the principal industrial source of vanadium used in steel alloys and for other purposes. Vanadium occurs diffusely in sea water in almost undetectable amount, but the larger sea squirts somehow manage to extract it and to concentrate it in their blood cells to an astonishing degree. No other creature or plant does it to such an extent. Sea squirts are not collected, however, as raw material for vanadium, but industrial vanadium is for the most part collected from the sooty deposits formed in the funnels of ships burning fuel oil from Venezuela. The oil beds of Venezuela, like all oil beds, are of living marine origin, but unlike others they appear to have included the remains of untold numbers of generations of sea squirts which contributed their vanadium quotas to the general deposit. Since sea squirts have existed as such for at least half a billion years there has been time enough for this to happen.

925. How are sea squirts related to fish? As adults the nature and position of their gills and heart and small brain suggest relationship, but the character of their minute swimming larvae clinches it. The larvae possess a locomotory tail which has a stiff internal supporting rod called the notochord, equivalent to the backbone of higher forms (even the human embryo has its notochord when at a very early stage), and it also has lateral muscles comparable to those of fish, and a tubular nerve cord on the topside of the notochord which expands in front as a brain. All these features are typically characteristic of vertebrate animals as a whole, and sea squirts must be admitted to kinship as primitive or as degenerate relatives.

926. What are sea squirts' closest relatives? While their connection with fish and other backboned animals is unmistakable, they are even more closely related to the lancelet (see question 906). They have very close relatives, however, which are the selps and pyrosomas (the name means fire bodies), transparent organisms present in enormous numbers drifting in the ocean currents.

FISHES

927. Where are fish likely to be found along the shore? Those fish that remain along the shore when the tide is out must hide as well as they can until the water returns, and inasmuch as they cannot do without some water they must find places which are both wet and safe. Rock pools at various levels are commonly frequented, especially those with seaweed in them; puddles lying beneath rock boulders are always worth examining; while shallow pools left among sand or mud flats by the ebbing tide often contain fish that were trapped before they could follow the receding water. The most likely places, however, are at the water's edge among the weedy rock crevices or among beds of eel grass and turtle grass; while some fish burrow in wet sand or take refuge in the burrows of other creatures. (See question 41.)

928. What fish are likely to be found on the shore or at the water's edge? Mostly slow-moving kinds that tend to keep to one locality or fish caught in pools when the tide recedes. The following are most likely to be found: rock eel, sculpin, cabezon, midshipman, moray eel, flounder, skate, grunion, sand eel, toadfish, lumpsucker, pipefish, sea horse, Cleveland goby, blind goby, blenny.

929. Where do sea horses live? The northern sea horse (*Hippocampus*) is known from Charleston to Cape Cod along the Atlantic Coast, where it may be locally abundant among weeds and sheltered bays during summer months. In winter it disappears, but where it goes is a mystery. Along the Gulf Coast of Florida as far as Pensacola Bay the dwarf sea horse is generally found living among the eelgrass or turtle grass in shallow lagoons. Sea horses are not known from the Pacific Coast of North America.

930. Can sea horses swim? After a fashion and as a rule in an upright position. By using the pair of small, ear-like pectoral fins just behind the head, together with the dorsal fin, they can glide very slowly and gently through the water, sufficiently well to move from one part of a weed bed to another but not well enough to contend with a strong current should they lose control.

931. What is the curved tail of the sea horse used for? For holding on to narrow blades of turtle grass or seaweeds, which they do most of the time.

932. What are sea horses like when they are born? They are more like pipefish (see question 938) than sea horses, and usually swim in a horizontal position by means of the tail which, while prehensile like their parents', has a tail fin more or less like that of other fish.

933. How do sea horses feed? They remain almost motionless attached to weeds. Whenever small crustaceans or equally small creatures swim or drift by their head, the sides of the head inflate and water and the victim are suddenly drawn in through the small, round mouth.

934. Why is the body of the sea horse so hard? The body is encased in armor which is made up of interlocking bony plates. Where the plates join together they protrude in ridges, knobs or spines. The armor is probably protection against crustaceans and other predators

Seahorse (1–4″)

that also live in the jungle of weed, although how necessary it may be is uncertain. The peculiar hard, jointed nature, however, may be partly a camouflaging effect and partly a means of acquiring a prehensile tail.

935. How do sea horses breed? In essentially the same manner as pipefish. The male and female mate at the time of egg expulsion, and the eggs are fertilized as they are shed. The female deposits them, however, on the abdomen of the male, whose pelvic fins have become converted into a large incubating pouch for holding the eggs. A brooding male sea horse thus has a pregnant look which the female lacks.

936. How large do sea horses grow? Most of them, including the northern sea horse, reach a length of five to six inches when fully grown, but the dwarf sea horse is never more than two inches.

937. Where can you find pipefish? Among the beds of turtle grass and eelgrass along the Atlantic Coast, especially among the Keys and Gulf Coast of Florida, and in particular in the bays and creeks, and among the eelgrass regions of southern California, sometimes as far north as San Francisco.

938. How can you recognize a pipefish? Only by watching the blades of turtle grass very closely, for the pipefish looks remarkably like a blade of grass itself, being unbelievably long and slender and more

Pipefish (5–15″) holding onto turtle grass

often than not remaining upright and as still as the weeds. Look long enough and you may see the greenish grass-blade body tilt forward and slowly swim, rooting with its long bony snout at the base of the grass.

939. How do pipefish feed? They search slowly and carefully for small swimming creatures, especially crustaceans, that live among the submerged jungles of turtle grass or eelgrass. Each small creature is examined carefully before being swallowed. If it is suitable, the back of the mouth is suddenly inflated and water rushes in through the narrow mouth itself, carrying the small victim inside.

940. Are pipefish related to sea horses? Pipefish (*Sygnathus*) are closely related to the sea horses, and in fact a newly hatched sea horse looks more like a pipefish than its own parent. The pipefish body is longer and straighter than that of the sea horse and has a tail for swimming instead of for holding on to weeds. (See question 932.)

941. How do pipefish breed? They are very similar to seahorses in their breeding habits. During the process of mating the eggs, as they are fertilized, are transferred to the safe keeping of the male. They adhere in rows along the underside of the body and are protected by two folds of skin which meet to enclose the eggs. Even after hatching the young fish remain close to their father and may re-enter the pouch for protection. (See question 935.)

942. What are small fish often found fastened to rocks, weeds or the underside of floats? On the Atlantic Coast the young of the lump-sucker (*Cyclopterus lumpus*), which is a fairly large, lumpish, sluggish rock-bottom fish, are often found attached to the undersides of floats and weeds. They hold on by means of their sucker-like ventral fins,

Lumpfish (1–20″) and clingfish (2–3″)

while the vivid green body and golden-flecked eyes of the young give them an enchanting appearance. On the Pacific Coast a small unre-lated fish with similar propensities is the clingfish (*Caularchus mean-dricus*). It is very common between the tides, either slithering over smooth wet surfaces of rock and weed or else clinging fast by means of its ventral sucker.

943. How do lumpsuckers hold onto weed, rock or wood surfaces? Their ventral fins are fused together to form a large sucker with which they can clamp on or let go at will.

944. What is a toadfish? A large-headed fish (*Opsanus*) somewhat like a sculpin, up to a foot long, which ranges from Maine to Cuba but is most common from Cape Cod to Cape Hatteras.

945. Where does the toadfish live? It is generally found in shallow water, hiding among weeds or rubbish, a habit which when put with its ugly appearance is responsible for its name, especially considering its sluggish nature and the fact that it hibernates in mud during winter.

946. What do toadfish feed on? Mainly on crustaceans, mollusks, small fish and any creatures of the right size which are careless enough to come too close to the trap-like mouth.

947. Are toadfish dangerous? Not if you leave them alone. But the toadfish is very pugnacious, especially when guarding its eggs, and is capable of inflicting very nasty bites when interfered with.

948. Where do toadfish lay their eggs? Often in rock crevices, very frequently in empty but still-imbedded pen shells, and, now that mankind tosses its trash into the sea, in almost any old shoe or empty tin can lying around in shallow water at a below-low-tide level. One or both parents guard the eggs. Spawning takes place in June or July.

949. What are the eggs of the toadfish like? The eggs are unusually large, being close to one-quarter inch across. They are also adhesive and stick to the wall of the shelter. Depending on the temperature, they hatch out after ten days to a month from the time they are laid. The developing fish are most interesting to watch if kept in a dish of seawater, in a cool place; the water should be changed at least twice a day. Remember, however, that a toadfish guarding its eggs is a formidable creature. (See question 947.)

950. What are sand eels? Sand eels (*Ammodytes*) are not eels at all but are slender sharp-pointed silvery fish four to eight inches long that swim in great schools in the coastal waters on both sides of the North Atlantic and are much preyed upon by both porpoises and sea birds.

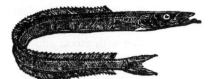

Sand eel (4–6")

951. Why are sand eels so named? Any long and slender fish is likely to be called an eel, but these are known as sand eels because they dive for the sea floor and burrow into the sand like a flash of lightning when under attack. There are no projecting fins to impede progress, while the shovel shape of the lower jaw aids them in entering the sand. A whole school may slide to safety in this manner beneath a few inches of soft sand.

952. Where can you find sand eels? They are often seen swimming in large schools in shallow water above sand flats, but if you dig in the clear sand exposed at the edge of the water at low tide, particularly where the sand lies in long ripples made by the undertow of the ebbing tide, you may dig them up in large numbers—or you may not. Even if you do, you need to be quick to catch them, for they can slip back into the sand about as fast as you can dig them up.

953. Where are grunion found? The grunion, a smelt-like fish (*Leuresthes*) about six inches long, swims in schools along the coast of

Grunion (5–7")

southern California and at certain times comes on to sandy beaches to deposit eggs.

954. When do grunion spawn? Only at night at the highest tides, generally between 9 P. M. and midnight, at the dark of the moon and at full moon when the tides are higher than at other times, during the months of March, April, May and June—just after the turn of the tide, when the tide begins to ebb.

955. How do grunion know when the tide is right for spawning? Presumably they don't! Nor can the light of the moon be entirely responsible, since spawning occurs both at full moon and at the dark of the moon. Eggs ripen every two weeks during the first spring and summer, but what sets the spawning clock in time with moon and tide has yet to be discovered.

956. How does the grunion lay its eggs? Grunion come to shore just after the high tide has turned and allow themselves to be carried up the beach with the surf. Usually they come in pairs, male and female, although as many as three males may come with a single female. They go up the beach as far as the waves take them and then the female digs into the sand tail first to lay her eggs, the male arching around her body to fertilize them as they are laid. With the wash of the next wave the fish slip back into the sea.

957. How long do grunion eggs remain in the sand? Almost exactly two weeks. They are laid during one spring tide, either at full moon or at the dark of the moon, and hatch at the time of the next spring tide two weeks later when the water floods the high sands again. The eggs hatch as the water reaches them and the larvae swim down to the sea at once.

958. How old are grunion when they breed? Grunion are mature at the end of one year and lay most of their eggs at that time. They may live to be three or four years old but only 25 per cent of the two-year-olds spawn and only 7 per cent of the three-year-olds.

959. How do you tell the age of a grunion? Grunions are scaly fish, and the scales show typical annual growth rings. If a large number of spawning fish are examined, the proportion of the one-, two- and three-year old groups is readily determined.

960. Are grunion good eating? So much so that the grunion spawning time once was as much a festival along the beaches of Lower California as the annual palolo-worm appearance in the southwest Pacific. Beaches used to swarm with hungry humans, and thousands of beach fires lit up the scene as the fish were roasted fresh from the sea.

961. Are the grunion under protection? Yes. So many were taken in the past, just at the moment of spawning, that the fish were in great danger of extermination. Now there are very strict laws limiting the taking of grunion.

962. How do flounders make themselves invisible on the sea bottom? By two means. Flounders (*Pseudopleuronectes*) can alter the pigmentation of the skin of the upper side so that it takes on the same general tone as the background. This process however takes a little time, since various kinds of pigment cells in the skin have to expand and contract to get any particular result. A flounder which disappears after a quick dash does so by making a shivering movement that throws mud or sand over its back, leaving only its eyes and mouth uncovered.

963. What do flounders eat? They wait for small fish to pass overhead, then they make a quick upward dash to capture them. At other times flounders move around, nosing in the mud or sand for worms and small mollusks and crustaceans.

964. Where can you see flounders? By wading around in shallow water or mud flats or on sandy mud. Usually you don't see a flounder until it darts away. Then if you carefully follow the streak made in the mud to where it stops, you may be able to step on the fish and pick it up with your hands. The chances are, however, that you will step a few inches off the mark and will see the fish dart away again. If you have nothing else to do this can be a pleasant, time-consuming occupation.

965. Is the darker, uppermost surface of a flounder the back of the fish? No. You are looking at its left side. The fish lies with its right side on the ground. The right side is a dirty white and only the left side is colored. Both eyes are on the left side of the face.

966. Are flounders born with two eyes on the left side of the face? No. When flounders hatch from the egg they look like typical fish, with one eye on each side of the head. After two weeks swimming in the upper water, by which time they are about half an inch long, the bridge of the nose dissolves and the eye of the right side slowly migrates across and comes to rest on the left side. When this is com-

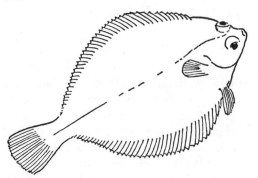

Flounder (8–16″)

pleted the fish sinks to the bottom and spends the rest of its life lying on its right side.

967. Can flounders use their eyes independently? Yes. If you watch one carefully, either on the mud or in an aquarium, you will see that each eye moves around almost as though it were mounted in a turret, and independently of the other.

968. Where can the common skate be found? The common skate (*Raia*) is frequently seen in shallow water on sandy or gravelly bottoms from Maine to Long Island Sound. It usually keeps away from the shoreline but is often left in shallow pools by the retreating tide, particularly where the tidal ebb is extensive.

969. What are the two holes immediately behind the eyes? They are the first pair of gill slits, modified to allow water to be taken in to the throat for breathing purposes while the mouth itself is occupied

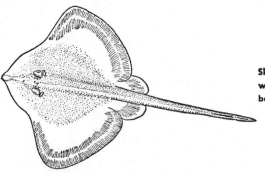

Skate (18–24″), showing water-intake passages behind the eyes

with feeding. They are called spiracles and correspond to your ear passages.

970. What kind of teeth does a skate have? They are flattened into closely fitting plates on each jaw so that the two jaws form a grinding mill capable of crushing the clams it feeds on.

971. What does the flat upper surface of the skate represent? It is the true back surface, unlike the upper surface of the flounder which is its left side, and the broad wings are the pectoral fins extending out on either side.

972. What are the five pairs of slits on the undersurface of the skate? They are the five pairs of gill openings, through which the water escapes that is taken in through the spiracles.

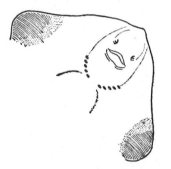

Undersurface of skate, showing mouth, nostrils and five pairs of gill passages

973. How do skates reproduce? The common skate usually lays two large eggs, each of them yellow and yolky, about the size of a pigeon's egg, and enclosed in a black, horny, four-pronged purse. The purses

Egg purse of skate (4″)

are more or less anchored to the sea floor. The embryo takes the better part of a year to develop and finally emerges from its case looking like a small edition of its parent. The empty cases are a common sight along the beaches. (See question 138.)

974. Where can the midshipman fish be found? The midshipman (*Porichthys*), also known as the grunt fish, grunters, or talking fish, lives on muddy bottoms of the ocean at about 300 feet depth along the Pacific Coast, but comes into shallow water and even tide pools during their breeding season. On the east coast it is found from South Carolina to Texas and is particularly common around Galveston.

975. Why is the midshipman fish so called? Because it has a row of golden spots down each side like the brass buttons on a midshipman's coat.

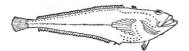

Midshipman (8″)

976. Why are midshipman fish also called grunters? Because they croak when disturbed, making a noise something like that of tree frogs. The sound is made by vibrating their air-bladder. In some regions they are called "singing fish."

977. Where and when do midshipman fish breed? They come into estuaries, bays and tide pools to spawn during June and July, at least in Newport Bay and Elkhorn Slough.

978. How do midshipman fish guard their eggs? A pair of fish dig a hole under a rock and the female attaches her rather large eggs to its undersurface. The male fertilizes them during that time and thereafter both parents remain on guard until the young are hatched. Until all the yolk of the egg has been absorbed, the young fish are unable to break loose from the rock, by which time they are about one inch long.

979. How did the cabezon get its name? Cabezon is Spanish for "big head," well suited to this large-mouthed, smooth-skinned fish (*Scorpaenichthys*).

980. Where is the cabezon found? In tide pools along the rocky shore of the Pacific Coast from Puget Sound to San Diego.

981. Do cabezons stay in the same pool all the time? Yes and no! Cabezons move up and down the shore to some extent as the tide comes in and goes out, and may inhabit pools at high tide which are dry at low tide. The same individual fish, however, may occupy the same place at a certain stage of the tide for days on end.

982. How large does the cabezon grow? The general run is between five and ten pounds, but they are known to grow as large as twenty-five pounds.

983. What do cabezons feed on? Any slow-moving animals likely to be found in tide pools and among the rocks below low tide, such as crabs, spiny lobsters, snails, octopi, worms, etc. One twelve-pound fish was taken with three fair-sized abalones in its stomach. They are also scavengers, for they are not fussy whether their food is dead or alive.

984. Can cabezons change color? Yes. They have a mottled color which makes them more or less inconspicuous in any case, but they can alter the mottling to match their surroundings so that they are virtually invisible except when they move. The camouflage is probably more a trap for the unwary than a necessary means of defence.

985. What are rock eels? The common rock eel (*Pholis*), also known as the gunnel or as the butterfish, is an eel-shaped fish, six to

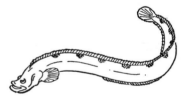

Gunnel or rock eel (5–8″)

twelve inches long, which slips through your fingers as though it were greased with butter.

986. Where can rock eels be found? Only on the Atlantic Coast south to Cape Cod. The fish may be found under almost every other rock boulder in the lower part of the shore, or among mussel beds, where there is a mixture of stones and some mud, together with a little water left by the retreating tide.

987. How do rock eels breed? They spawn during late winter or early spring months, laying their eggs in a large cluster beneath a rock or in

Rock eel guarding eggs in oyster shell

an empty shell. The male fish curls around the egg mass until the young fish hatch out.

988. Is the moray eel a real eel? Yes, although the moray eel (*Gymnothorax*) does not enter fresh-water rivers and lakes to feed. It hunts for its food along the rocky shores of the Pacific Coast and from southern Florida through the West Indies.

Moray eel (3–7′)

989. What does the moray eel usually feed on? Moray eels are specialists in the fine art of hunting and eating octopi. The fish are long and slender, although not so round as the fresh-water eel, and go down the dark narrow crevices between submerged rocks in search of octopi.

990. How does a moray eel overpower an octopus? "When the eel grasps the octopus in its mouth, the octopus immediately fastens to the head of the eel with its tentacles. The eel then throws a loop with its tail around its own body and slips its body backward through this loop, thus forcing the tentacles of the octopus loose from its head, and at the same time makes another gobble or two to get the octopus farther down its throat. This performance may be repeated several times before the octopus is finally swallowed. If the octopus is too large for the eel to swallow, it grasps a tentacle and then spins its body at full length, thereby quickly twisting the tentacle from the body of the octopus. The octopus may then be eaten an arm at a time, or it may escape with one arm missing."—(G. E. and N. MacGinitie.)

991. Is the moray eel dangerous? Yes. Many people have been bitten by moray eels while hunting for abalones along the coast of southern California. The eel is equipped with a formidable set of teeth which can inflict nasty wounds. It is dangerous to reach into crevices or under rock ledges, where the moving hand or fingers may be taken by the eel to be the tentacles of an octopus and treated accordingly.

992. Where are sculpins found? On both the Atlantic and Pacific Coasts. Sculpins (*Myxocephalus*) have much the same shape as the cabezon but have a scaly skin, whereas the cabezon is smooth. Sculpins are less likely to be found in rock pools and are more likely to be seen among the weed in rock crevices at low tide.

993. How do sculpins feed? They lie in wait, their big mouths ready to open suddenly like a trap. When the mouth opens, water rushes in together with any creature caught in the swirl. The sculpin head is

Sculpin (9–12″)

decorated with folds of skin and various projections, while the body is mottled dark and light, so that the fish is most inconspicuous as it lies among the rock and weed.

994. Can sculpins change color? Yes. The color pattern remains more or less the same, but it can be made lighter and darker according to circumstances.

995. Are the spines of the sculpin dangerous? They are not actually dangerous, but they can cause an extremely painful wound, and sculpins are best not handled. The sting of the Pacific Coast sculpin, more properly known as the sea scorpion (*Scorpaena*), is poisonous and causes a large swelling with severe pain lasting for about one hour and the swelling much longer.

996. What does the spawn of the Pacific sculpin or sea scorpion look like? The spawn is peculiar and is produced several times during late summer. It consists of two pear-shaped balloons of jelly containing the eggs, the two balloons being joined together at their narrow end. After being spawned, the double balloon floats to the surface and drifts away, the larvae hatching in due course.

997. Is there any advantage in spawn consisting of many eggs in a large jelly mass? There is both advantage and disadvantage. Having all eggs in one gelatinous mass is almost literally a case of putting all one's eggs in a single basket, and if anything goes wrong all of them are lost. On the other hand such jelly is generally unpalatable and in any case cannot be swallowed by the numerous small creatures that would otherwise be feeding on the individual fish eggs and embryos.

998. What fish lives in the burrows of ghost shrimps and innkeeper worms? The Cleveland goby (*Clevelandia*) is a transient guest in the burrows of the shrimp and the worm, foraging outside but using the burrows as refuges, while the blind goby (*Typhlogobius*) is a permanent resident of one of the southern California ghost shrimps. (See question 611.)

999. Where do Cleveland gobies live? This little mud or sand colored fish, one to two inches long, is the commonest between-tide fish along the whole West Coast from Vancouver Island to Lower California. They are everywhere except on the exposed mud flats, gathering in tide pools and water channels, among the roots of eel grass and in the burrows of most any animal in the neighborhood. Anywhere from one to twenty-four Clevelands have been found within a single innkeeper's burrow when the tide was out, while the MacGinities once collected over 400 in a shallow tide pool about two square yards in area.

1000. Why are Cleveland gobies so hard to see? Because they match the color of their skin to that of the sand and mud. Unless they are moving they blend so perfectly that they cannot be seen.

1001. What do Cleveland gobies eat? Principally the minute crustaceans and larvae of worms, clams, etc., that continually settle on the mud flats and elsewhere. They can swallow only the tiniest morsels.

1002. What happens when a Cleveland goby finds food too large for it to swallow? "If a Clevelandia finds a morsel of food that is too large for it to swallow, with true Tom Sawyerish propensity it carries the piece of food to some crustacean, and, as the latter tears the food to pieces, the fish snatches particles to eat, and at intervals, even snatches the larger piece and attempts to swallow it. If it is still too large it is left for the crustacean to tear apart further."—(G. E. and N. MacGinitie.)

1003. What feeds on Cleveland gobies? In spite of their invisibility when still, the Cleveland gobies are preyed upon extensively. Among the birds, the willits, godwits and curlews capture many by inserting their bills into the various burrows where the fish are hiding. Pistol shrimps stun them as they pass or enter the burrows in which they themselves are hiding. Their chief enemy, however, is the large-mouthed mud-flat fish (*Leptocottus*). (See question 603.)

1004. Where do blind gobies live? They live in pairs within the burrow of a ghost shrimp and apparently nowhere else (see question 611). Once within a burrow, the fish never voluntarily leave.

1005. How do blind gobies get enough oxygen inside a ghost-shrimp burrow? Partly by reducing the need for oxygen, for the fish are very sluggish. Furthermore, they no longer use their gills except when sudden exertion is required, but obtain oxygen mainly through the skin. By one means and the other they can live in water so stagnant that it would suffocate almost any other fish.

1006. What do blind gobies feed on? Chiefly small pieces of seaweed which happen to enter the shrimp burrow, although bits of animal material are readily eaten when available. Sorting of food is done by touch and taste, for even if the fish were not blind they would have little or no light to see by.

1007. Do blind gobies spawn inside the burrow? Yes. Depending on the size of the female she lays anywhere from three to fifteen thousand eggs and the male fertilizes them in batches as they are laid. The eggs are glued to the wall of the burrow wherever the surface is hard enough.

1008. Are blind gobies blind from the start? No. The newly hatched fish have perfectly good eyes and swim away from their parents' adopted home. Nothing is known about them until they become established as pairs mated for life inside a ghost shrimp burrow, when they are about six months old. By then membranes have already grown over their eyes. The growth of flesh continues until in larger fish the eyes may be completely hidden.

**Butterfly blennies (3"),
one in empty shell guarding eggs**

SEA TURTLES

1009. What is the difference between a sea turtle and any other kind? Sea turtles have flippers for swimming that look more like the fins of fish than the walking legs of a land animal, whereas even in the fresh-water turtles the legs are still obviously legs ending in toes despite their webbed appearance. Sea turtles have gone further in their adaptation to an aquatic existence, although they are just as truly reptiles and must come to the surface to breathe, and to the land to lay eggs.

1010. What kinds of sea turtle can you expect to see along the Atlantic and Gulf Coasts? Green turtle, loggerhead turtle, and the hawksbill turtle. They used to be common but those days are long

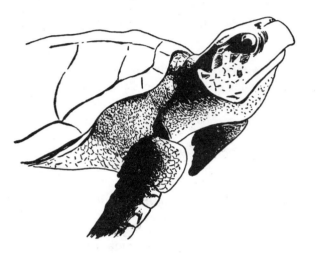

Green turtle (3–6')

since passed, for few wild animals can tolerate having their eggs used by bakeries and their flesh converted to soups and steaks—striking at both ends leaves little in between.

1011. Are sea turtles protected? Yes, but poachers still take such eggs and adults as they can find.

1012. How big do sea turtles grow? Green turtles rarely weigh more than 150 pounds where they are molested by man, but are known to attain to 850 pounds where left alone. Loggerheads rarely reach more than 150 under any circumstances.

1013. Is the flipper of a sea turtle a fin or a leg? It looks like a fin and is used as a fin, but it is made out of a leg such as that which the sea turtle's less adventurous relatives, the fresh-water terrapin and desert tortoise, still use in the manner for which it was designed. Yet if you take away the scales and flesh from the flipper, the bones of the skeleton are still clearly recognizable as corresponding to the bones in your own arm or leg.

1014. How long does a sea turtle stay under water without coming up to breathe? Usually from thirty to forty minutes.

1015. How can you tell the sex of a sea turtle? The males have a long tail, the females do not. Males, however, rarely leave the water and are seldom caught.

1016. Do sea turtles have teeth? No turtle has teeth, whether it be a sea turtle or a tortoise or the fresh-water terrapin. The teeth were lost long ago in the course of evolution, and the jaws are covered only by a horny but effective beak.

1017. What do sea turtles feed on? Loggerheads and hawksbills are omnivorous feeders and feed not only on hermit crabs, conchs and other shellfish, but also on jellyfish and even the Portuguese man-of-war. The turtle's strong beak and jaws easily crush shells, while the scales protect the animal from the stings of the man-of-war, although it has to close its eyes while feeding on them. The green turtle also feeds on mollusks and crustaceans, but in addition consumes large quantities of turtle grass.

1018. Where do sea turtles make their nests? In the upper part of sandy beaches above the level to which seawater can percolate even at the highest tides. If the salt water should reach the eggs the embryos within would die. The nest is simply a deep pit dug in the sand with the flippers and subsequently filled in and smoothed over.

1019. How many eggs are laid in a nest? In the case of the loggerhead, which is the most likely to be seen along the Atlantic Coast, even as far north as Cape Cod, the eggs in the nest usually number between 100 and 150.

1020. How long do sea-turtle eggs take to hatch? Loggerhead eggs and probably those of most other sea turtles take about two months to hatch.

1021. How do young, newly hatched sea turtles find their way to the sea? They tend to take a downward path in any case, but also head for any wide blue horizon, which is usually the right way.

SEA MAMMALS

1022. What is a porpoise? The common porpoise is one of the dolphin family, which consists of small-toothed whales. Porpoises are true warm-blooded, air-breathing mammals that bear their young alive and suckle them just as do all their land-dwelling relatives. The dolphin fish, however, is a large tropical fish somewhat resembling the dolphin in appearance and was consequently named after it, with some resulting confusion.

1023. What are dolphins? If we exclude the dolphin fish, which is a true fish and has no right to the name dolphin, the dolphins as a group are small-toothed whales and are the most voracious creatures in the sea. They include not only the smaller forms actually called dolphins, together with porpoises of similar size, but also larger kinds known as blackfish, the white whale or beluga, the narwhal, and the most ferocious of all sea animals, the killer whale.

1024. What is the difference between a porpoise and a dolphin? The porpoise and dolphin are both very small-toothed whales belonging to the dolphin family, the porpoise being distinguished from other dolphins of the same size by having a more or less dome-shaped forehead, made up of fatty tissue, whereas the common bottle-nosed dolphin, for instance, has a concavity between skull and jaw. The larger members of the dolphin family go by other names. (See questions 1023, 1031.)

1025. Where can porpoises be seen? In most parts of the ocean where fish are plentiful, and frequently in coastal creeks and tidal waters. They are particularly well seen on display at Marineland, Florida.

1026. Do porpoises always travel in schools? As a rule, at least several keep together and a lone porpoise is rare. Large schools of

School of porpoise (6–7′)

porpoises are common, as are those of other members of the dolphin family such as the killer whale, black fish, white whale and the little dolphins themselves.

1027. How intelligent are porpoises? At least as intelligent as an intelligent dog and possibly more so. They have a considerable vocabulary of sounds for intercommunication and possibly for sonar navigation under water.

1028. What enemies do porpoises have? Their chief enemy is the largest of all dolphins, the killer whale, which is the most ferocious and possibly the most intelligent animal in the sea. Killer whales may be as large as thirty feet and may be seen close inshore in Alaskan waters. Even a half-grown killer, sixteen feet long, when opened up was found to have swallowed thirteen porpoises and fourteen seals for its last meal.

1029. How often must a porpoise come to the surface to breathe? As a rule, every few minutes.

1030. How fast can porpoises swim? Porpoises and dolphins are capable of swimming thirty miles an hour without apparent difficulty.

1031. What whales are likely to be seen from the shore? Blackfish, which are large dolphin whales about fifteen feet long, occasionally run aground on sand bars and beaches as they drive through the water after fish, and become stranded as the tide falls. As a rule this is fatal, because without the support of water their own weight prevents them from breathing adequately. The white whale or beluga is about the same size and may be seen from cliffs or ships along the St. Lawrence River, particularly near the entrance to the Saguenay River. As many as 800 have been seen swimming together. Both the blackfish and white whales have teeth and feed on fish. Neither is noticeable at the surface when it comes up to breathe.

In contrast to these, you can often see two whalebone whales, as they come to the surface to blow. These are whales without teeth, which strain small crustaceans from the water as it is passed through whalebone sieves in the mouth. They are the fin whales and humpback whale, whales that are much larger than the dolphin family although small for their own class, running about fifty or sixty feet long. They are most likely to be seen from the shore in colder waters— that is, north of Cape Cod on the Atlantic Coast and along much of the Pacific Coast. Even a mile away, the spume sent up when the animal surfaces can be seen from the shore. The spume or "blow" rises about twelve feet into the air. If nothing else is seen, the whale is probably a fin whale perhaps as long as sixty-five feet. If the broad flukes of the tail come up into view as the whale dives, it is most likely a humpback and may be fifty feet long.

1032. What kinds of seals can be seen from the shore? The harbor seal is the most common and is most likely to be seen along the inhabited coasts of North America. The California sea lion, which has a pair of short external ears in contrast to the harbor seal which has none, lives only along the Pacific Coast. This is the performing seal that you see in the circus or on the stage. The other seals of the North Atlantic and North Pacific are beyond the range of the average beachcomber.

1033. Where can harbor seals be seen? Harbor seals extend along the Atlantic Coast from southern New England to the Arctic Ocean and from Lower California to northwest Alaska. They may be seen most readily in the bays and estuaries and on rocky islets close to shore. The California sea lion can be seen in summer near Monterey and San Francisco.

1034. Are seals protected? The sea lion is protected. The harbor seal is protected in New England waters but not in the Pacific, nor in Canadian Atlantic waters where it has a bounty on its head for being a suspected carrier or transmitter of codworm.

1035. What do seals feed on? The California sea lion lives chiefly on squid, contrary to fishermen's belief that they destroy great numbers of food fish. Harbor seals feed on a large variety of fish, most of which are bottom forms of little importance to fishermen, with squid included when available.

1036. How long can they stay under water? For a good many minutes if necessary but usually for not more than five. The harp or Greenland seal can stay under for twenty minutes without coming up to breathe.

1037. Where do seals breed? California sea lions have breeding colonies on the "Seal Rocks" of San Francisco and the rocky islands off Point Lobos near Monterey. Harbor seals bear their young in the river estuaries where the right food is to be had when weaning time comes.

1038. Do the male harbor seals go into the estuaries with the pregnant females? No. They remain on the offshore islets, fishing and resting to their hearts' content.

1039. Can they swim when they are born? No. Young seals have to learn to swim, with aid from their mother. When tired, the pup is usually given a lift on the mother's back.

1040. How long do they nurse? From four to six weeks.

1041. How intelligent are seals? Among the most intelligent of all animals—monkeys and men excepted. The California sea lion is especially smart and is also fun-loving and affectionate. The harbor seal is notable for its great curiosity.

1042. Is the fur of the harbor seal used commercially? Not to any extent. The coat is hairy rather than furry. The other name for the harbor seal is the hair seal.

1043. How common are sea otters? At one time they were present by the thousands along the whole length of the North Pacific Coast, but fur hunters brought them near to extinction. Now heavily protected, they are making a comeback and several thousands are known to exist among the Aleutian Islands.

1044. Where can sea otters be seen? With the aid of a telescope you may be able to see a few sea otters off shore in the kelp zone off Point Lobos near Monterey. Otherwise you may take a trip to the Aleutians.

1045. What do they eat? In the north they feed on sea urchins mainly, with practically every other kind of sea creature eaten to vary the diet. Along the California coast they feed on red abalones, sea urchins and crabs. (See question 351.)

1046. Where do sea otters sleep? "The sea otter herd goes to bed with the coming of darkness. A favorite dormitory of kelp is used night after night. Ribbon-like kelp is preferred to the tubular species because it is easier to wrap up in and is less likely to slip loose."— (V. H. Cahalane.)

1047. Where do they breed? After a nine-months pregnancy a single pup is born on an isolated rocky islet or on a thick bed of kelp in a sheltered cove. The pup is nursed until it is about one year old, although crab and shellfish meat is fed to it from about six months on. If the pup gets tired of swimming lessons, the mother lets it ride on her chest or brings it in her arms back to the kelp bed.

1048. Where does the manatee live? A few manatees probably still browse sleepily in the lagoons of the Florida Thousand Islands and Florida Bay, while others drift about in Biscayne Bay and north to Fort Lauderdale. They are still numerous in wilder parts of Central and South America.

1049. What is a manatee? A manatee is a mammal and an aquatic one at that, at least as much so as a seal though of a very different class. It is also the original mermaid, in spite of its obvious ugliness, and was described as such by Columbus on his second voyage to the Caribbean. Sometimes called sea cows, they may weigh as much as a ton.

Manatee family
(9–15')

1050. What does a manatee feed on? They eat masses of slender river grass, sitting upright with their heads above water, reaching for the grass with their fingers. The rest of the time they doze in the water, rising to the surface to breathe every five to fifteen minutes.

1051. How does it reproduce? "In March or April the manatee mother looks for a sheltered place that is safe from sharks, killer whales and sawfish. It may be a lagoon—provided it is not a crocodile resort—or behind a mass of driftwood or wreckage. This is the birthplace of the little manatee (occasionally twins). About thirty inches long, it can—it must—swim immediately. Being born in the water, this is one of the few mammal young that needs no swimming lessons. The mother manatee nurses her young by sitting up, her head and shoulders out of the water, and clasping it to her breast between her flipper-arms—a Madonna of the Sea!"—(V. H. Cahalane.)

SHORE BIRDS

1052. What birds dive into the sea from the air? Gannets, boobies, terns, brown pelicans, ospreys and kingfishers. Gannets and boobies dive like arrows shot into the water, penetrating to some depth. The brown pelican also drops from a height but barely dives out of sight. The white pelican does not dive (see question 1085). Terns dive in good style but are too light in body to plunge more than a few inches beneath the surface, a statement which applies to the kingfisher, too. The osprey drops onto the water feet first with talons extended to seize a surface-swimming fish.

1053. What birds dive only from the surface of the water? Cormorants and the Pacific loon, and diving ducks in season. The cormorants, especially, are marine fishermen, adept at swimming underwater and catching fish, but always alighting on the surface of the water first and diving from a sitting position.

1054. What birds other than gulls and herons are likely to be found along the shore? The smaller wading birds such as sandpipers, sanderlings, willits, godwits, knots, curlews, yellowlegs, turnstones, oyster catchers, avocets, and various plover including the semi-palmated, piping, blackbellied, snowy and killdeer, all feed along the shore line. Skimmers, terns, pelicans and ospreys dip or dive into the shallow water from the air, more concerned with fish than with crabs or snails. Chunky, blackish scoters are often seen flying in low formation over the waves in the more sheltered bays, or busy diving from the surface in shallow water, as do the cormorants and loons. Wild geese—the American brant—winters along the Atlantic Coast and may be seen feeding on the mud flats in secluded bays. Both the Pacific and red-throated loon also winter along the Atlantic Coast.

1055. What do shore birds feed on? "Sandpipers search the wrack for food and run up and down the beach with the shore waves, picking up anything edible. Along the shore their food consists of sand hoppers, mysids, isopods, worms, etc. A flock of sandpipers wheel, turn, or otherwise change direction as one bird. Sanderlings have habits quite common to those of sandpipers. In addition they have an uncanny ability to find grunion eggs that are near the surface. Once the eggs are located, a hundred or more sanderlings may collect at the feast, peeping away for all the world alike to huddles of baby chicks feeding. Willits and godwits feed along the beach, shoving their long beaks deeply into the sand for sand crabs or worms. Turnstones feed mostly on rocky beaches, turning over small stones to pick up the worms or crustaceans hidden under them . . . Curlews feed either

Avocet (16–20″)

Turnstones (9″)

Oyster catchers (17–21″)

Ringed plover (6″) and sandpipers (7″)

Curlew (2′)

along shores of inlets or on sandy or rocky shores of the ocean beach."—(G. E. and N. MacGinitie.)

1056. What do shore birds have in common? Relatively long legs with specialized feet, and a long bill.

1057. What birds fish from the shore? The herons, which include the blue heron, the green night heron, the American egret and the snowy egret. Wood ibis and white ibis should also be listed with those that feed among the mangroves of the southern coast.

1058. Where do the herons and egrets fish? Standing at the edge or in a few inches of water along the shore, whether the shore be rocky, or mud flats, or mangroves, particularly when the tide is low and fish are more likely to be discovered.

Heron (3–4')

1059. How do they fish? Mostly they stand perfectly still, their long legs and necks giving them height enough to see well into the water to spot a fish, their long legs making it possible to wade, with the long flexible neck and long spear-like beak ready to strike into the water like a thrown javelin. "When fishing, the snowy egret places a foot down in the mud ahead and shuffles its foot rapidly to stir up any hiding *Clevelandia, Leptocottus, Gillichthys,* or small flounders. As one of these fish scoots away, the bird watches carefully where it stops and very slowly proceeds to the spot to spear the prey. We have never seen the great blue heron or American egret use its feet in the same way as the snowy egret.

"The only night-fishing bird of the mud flats that we know is the green night heron. It has been our companion many nights as we studied the nocturnal habits of mud-flat animals, but darkness prevented our seeing much of its activities. We could hear it splashing and squawking, and occasionally one would go by overhead and we could hear the soft swishing of its wings."—(G. E. and N. MacGinitie.) (See questions 998, 962.)

1060. How far offshore do gulls fly? As a rule not very far, for they are essentially birds of the seacoast. A flock of gulls following an outgoing ship usually melts away as the land is left behind. In the North Atlantic, however, kittiwake gulls may follow a ship right across from land to land.

1061. What do gulls feed on? On practically any kind of flesh available, dead or alive. Primarily they are scavengers, feeding on refuse and dead animals from the surface and coastal waters. Consequently they play an important part in reducing pollution in harbors. They also hunt along the shore for crabs and mollusks—and they are robbers that take the eggs and even the chicks of other birds that nest along the coast, particularly those of terns and cormorants.

1062. Do gulls dive? No. At the best they make a hasty landing on the water where fish are breaking surface, but they cannot dive in the sense that terns, gannets, pelicans or even cormorants do—neither from the air nor from the surface of the water.

1063. Where do gulls breed? Those that nest along the seacoast form nesting colonies on the uninhabited islands and more remote rocky headlands where they can do so unmolested by egg-stealing mammals, human or otherwise. Nests are usually on the ground and are simply made, with two or three eggs laid as a rule.

1064. What do the young gulls look like? Nestlings are generally covered with mottled down of yellow, brown, grey or black. Young but fully grown gulls are dusky grey or brown, very different from the white and black of the mature adult.

1065. When do young gulls come to the shore to feed? Immature but fully grown herring gulls and black-backed gulls come ashore with the parent gulls usually early in August, and may be seen following the mother along the shore keening and crying for the food she finds. Gradually the young gull is left more and more to find its own food and by September has been fully trained to take care of itself.

1066. What gulls are common along the Atlantic and Pacific Coasts? The herring gull, laughing gull, and the great black-backed

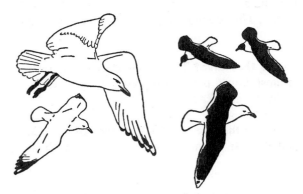

Herring gulls (left—2′), laughing gulls (right top—16″),
and great black-backed gulls (right lower—30″)

gull are common along the Atlantic Coast, while the California gull, glaucous gull and western gull are the common gulls of the Pacific Coast.

1067. How can you tell a tern from a gull? The terns of the Atlantic and Pacific Coast are generally much smaller than even the small-

Terns (14″)

est gull and can be distinguished by the presence of a strongly forked
tail, which gulls do not have. Moreover, they have an easily recognizable dipping, swallow-like flight. When flying over the sea they
continually scan the surface and the beak is usually pointed downward
at right angles to the body.

1068. What do terns feed on? Mainly on small fish near the surface
of the water, but small squids, crustaceans and mollusks are also taken
when available. They do not feed on dead fish or floating refuse of any
kind.

1069. How do they dive? They circle over the water where the fish
are and almost hover before dropping straight down from ten to
twenty feet in the air, head first.

1070. Where can they be seen? Along the Atlantic Coast and parts
of the Pacific Coast, usually where small fish, such as herring, are
abundant.

1071. Where do they breed? The common, royal and least terns of
the North American Coasts breed in colonies on secluded sand bars
above high tide.

1072. How can you tell a cormorant from a duck? There is no difficulty if you can get a close view, for a cormorant has a slender hook-tipped beak very different from a duck's bill, and has its very short
legs placed far back on the body. In the water the cormorant sits much
deeper than a duck, with only its head and neck visible above the surface, much like a loon. Cormorants are much larger than any ducks,
and in flight can be easily distinguished by this alone. Cormorants are
black underneath, where loons are white.

1073. What is the difference between a cormorant and a shag? On
this continent shag and cormorant are alternate names for the same
bird.

1074. How do cormorants dive? They dive from the surface of the
water by first springing a few inches into the air and then diving head
foremost, with wings closed.

1075. How do cormorants swim under water? They pursue their prey under water, using their large feet for swimming among rocks and seaweed, and both wings and feet when swimming in clear water over sandy or muddy bottoms.

1076. What do cormorants feed on? Chiefly on bottom fish and crustaceans.

1077. Why do cormorants stand on rocks or channel markers with wings outspread? The plumage of cormorants is by no means so resistant to water as that of many other diving birds, and after a certain

Cormorant (30–35″) drying wings

amount of diving tends to become waterlogged. The wings and body feathers accordingly have to be dried out every so often.

1078. Do the wings of the cormorant beat the water when the bird is rising from the surface after a dive? No. The succession of splashes that you see when a cormorant is taking flight from the sea surface comes from the use of the powerful feet. As the bird tries to take off by beating its wings, it simultaneously propels itself forward by running rapidly with its feet, the splashes from which at a distance look as though they were made by the downbeat of the wings.

1079. Do cormorants have to wet their tails before they can fly? They do not have to, but sometimes they cannot avoid doing so. When starting to fly from a rock or a marker close to the water they have to fly downward in order to get enough momentum, and often touch the water before they begin to rise.

1080. Do cormorants fly in formation? Most of the time cormorants fly singly or in pairs. Even then, they don't waste time but fly in a straight line from where they are to where they intend to go, without the meandering other birds are inclined to do. When a group of cormorants are in flight, they are generally in a line one behind the other.

1081. Do cormorants do any harm to fisheries? No. In spite of fishermen's opinions and complaints, cormorants feed principally on bottom fish of no commercial value, such as rockfish and weedfish. The food fish generally are too fast to be caught.

1082. Where do cormorants breed? They breed along the Atlantic and Pacific Coasts and also in the interior of the continent, but the small outlying uninhabited offshore islands are the best places to see the breeding colonies.

1083. Why do brown pelicans usually fly in formation? The position and spacing of birds in line is determined by the need of clear vision ahead for all of them, and perhaps by the nature of the air movements set up by their beating wings.

1084. How do brown pelicans dive? "Frequently the birds would hurtle from a height as great as twenty meters above the water, usually spiralling or twisting on their downward course so that they struck with their backs rather than with their breasts toward the surface. In other words they would be gliding upside down, with wings still half spread, at the moment before the plunge. Usually they quite disappear, even to the tip of their long wings. . . . The twisting descent is doubtless responsible for the fact that the bird turns some sort of somersault under water, and comes up heading in another direction from that of its diagonal dive. Emergence, in fact, finds the pelican facing the wind and ready for flight, while the plunge is usually, if not always, made with the breeze astern."—(R. C. Murphy.)

1085. How do white pelicans catch fish? White pelicans, which are much larger than the brown and appear to be unable to dive, catch their food on or near the surface by swimming or wading in shallow water. "I have often noticed the birds in flocks, in pairs, or alone, swimming on the water with partially opened wings, and head drawn

Brown pelican (4')

down and back, the bill just clearing the water, ready to strike and gobble up the prey within their reach. . . . Their favorite time for fishing on the seashore is during the incoming tide, as with it come the small fishes to feed upon the insects caught in the rise, and upon the low forms of life in the drift as it washes shoreward, the larger fish following in their wake, each from the smallest to the largest eagerly engaged in taking life in order to sustain life. All sea birds know this and the time of its coming well, and the white pelicans that have been patiently waiting in line along the beach, quietly move into the water, and glide smoothly out, so as not to frighten the life beneath, and at a suitable distance from the shore, form into a line in accordance with the sinuosities of the beach, each facing shoreward and waiting their leader's signal to start. When this is given, all is commotion; the birds rapidly striking the water with their wings, throwing it high above them, and plunging their heads in and out, fairly make the water foam, as they move on in an almost unbroken line, filling their pouches as they go. When satisfied with their catch, they wade and waddle into line again upon the beach, where they remain to rest, standing or sitting, as suits them best, then, if disturbed, they generally rise in a flock and circle for a long time high in the air."—(N. S. Goss.)

1086. How do brown pelicans catch fish? "As it strikes the water, the beak is opened sufficiently to 'span' the fish. At the moment the beak strikes the water, the curvature of both upper and lower mandibles causes them to come quickly together. At the same time the inrushing water, carrying the fish with it, swells out the pouch of the lower mandible. . . . The pelican then holds its beak, up to its eyes, under water, contracts the muscles of the pouch, and forces the water out the slits on either side between the mandibles, leaving the fish stranded inside. The head and beak are then raised nearly vertically, and the fish is swallowed by a series of four or five gobbles."—(G. E. and N. MacGinitie.)

1087. How much can a brown pelican's pouch hold? When fully expanded with water after a fish has just been caught, the pouch may hold two gallons or more.

1088. What do black skimmers look like? Like large terns with almost-black upper parts and white forehead and underparts, forked

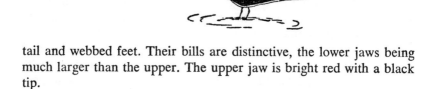

Black skimmer (18″)

tail and webbed feet. Their bills are distinctive, the lower jaws being much larger than the upper. The upper jaw is bright red with a black tip.

1089. Where are black skimmers seen? Along the Atlantic Coast of both North and South America, particularly where there are extensive sheltered shallow stretches of sea.

1090. When can they be seen? As a rule they are seen when skimming, which is done mostly early and late in the day and on moonlit nights. They usually rest on sandbanks during the daytime.

1091. How do they catch fish? They fly rapidly just over the water with the beak open and the long lower jaw cutting the water surface as though it were plowing the sea. Their food consists of anything thus caught, whether small fish, other animals or even weed.

1092. Where do they nest? They breed in colonies on sand banks, laying their several eggs in scooped-out hollows. Young chicks are sand-colored and almost invisible as long as they are still.

1093. How can you tell a gannet from a gull? They are larger, have a more soaring flight, and may be seen to dive, which a gull rarely does at all and never well.

Gannets (6' wing spread)

1094. Where are gannets to be seen? In the Gulf of St. Lawrence and off the coasts of Newfoundland, Labrador, and occasionally offshore along the New England coast, as far as New York City during the summer and even farther south in winter.

1095. How do gannets dive? Usually they dive from a height of about sixty feet, sometimes as high as a hundred, dropping like plummets and closing their wings at the last moment before they enter the water.

1096. How deep can a gannet dive? They have been caught in fishermen's nets as deep as ninety feet below the surface. When fish are deep the gannets plunge vertically; when the fish are close to the surface the plunge is oblique and the dive shallow.

1097. Where do gannets breed? On islands in the Gulf of St. Lawrence, the best known colony being on Bonaventure Island.

1098. Where can frigate birds or man-of-war birds be seen? The magnificent man-of-war or frigate is a tropical Atlantic bird which ranges as far north as the Florida Keys and the Louisiana Coast. It is almost certainly to be seen by anyone taking a trip to the most remote of the Keys, to see Fort Jefferson on the Dry Tortugas.

1099. How can they be recognized? The large black frigates have very long wings, long forked tails and very short legs. Given a close view they are the most readily recognized of all sea birds. The tail is not invariably spread, so that its forked nature is not always obvious. They usually fly high, soaring and coasting without effort, with motionless wings outspread.

1100. How did they get their names? From their marauding habit of forcing other sea birds to disgorge the fish they have caught. Man-of-war and frigate bird are names bestowed by the sailors of earlier days when they themselves were subject to piratical attacks from buccaneers.

1101. How do frigate birds obtain their food? They pick sea creatures such as fish, mollusks, jellyfish and shrimp from the surface of the sea with their long, hooked beaks while still in swift flight. Alternatively they swoop down from the sky and force boobies, pelicans, cormorants, terns and gulls to yield or throw up any fish they may have caught, giving the victim a savage bone-breaking peck if there is any hesitation in doing so.

1102. How do frigate birds compare in size and strength with other sea birds? They are second in size only to the albatross, but considering their somewhat smaller size they have a wing spread and wing power second to none—the most powerful flyers in the world.

1103. Do frigate birds ever land on the water? No, at least never by design. While their feet are webbed, they have long ago lost any habit of resting on the water. They are, in fact, so marvellously developed for flight that almost all else has been sacrificed and the bird is unable to rise from the sea.

1104. How does the osprey catch fish? It dives like a hawk, since a hawk it is, with feet and beak thrust forward to grasp its prey as it comes within reach. In other words, the osprey swoops down to the water, splashing into it with talons extended to seize the fish, which must be close to the surface. The bird does not enter or even properly alight on the water but takes off again at once with wings which were never folded.

1105. Where do ospreys build their nests? An osprey nest is always large and conspicuous as a solitary nest of sticks built at the top of a dead spruce tree, usually on a deserted island if such is available. Telephone poles are often used in place of spruce trees if suitably located. "At Pulpit Harbor on North Haven, Me., ospreys have nested continuously on Pulpit Rock ever since the first settlers arrived and in the memory of Indians as far as it goes—probably several hundred years at least."—(Robert F. Duncan.)

BOOKS

1106. What books can you get for further reading and for good photographic illustration?

Abbott, R. T. AMERICAN SEA SHELLS. D. Van Nostrand Co. 1954.
Alexander, W. B. BIRDS OF THE OCEAN. G. P. Putnam's Sons. 1955.
Berrill, Jacquelyn. WONDERS OF THE SEASHORE. Dodd, Mead & Co. 1951.
Berrill, N. J. THE LIVING TIDE. Dodd, Mead & Co. 1951.
Buchsbaum, R. ANIMALS WITHOUT BACKBONES. University of Chicago Press. 1948.
Cahalane, V. H. MAMMALS OF NORTH AMERICA. Macmillan Co. N.Y. 1947.
Carson, R. THE EDGE OF THE SEA. Houghton Mifflin Co. 1955.
Coker, R. E. THIS GREAT AND WIDE SEA. University of North Carolina Press. 1947.
Gruss, R. THE ART OF THE AQUALUNG. Philosophical Library, N.Y. 1956.

MacGinitie, G. E. and N. NATURAL HISTORY OF MARINE ANIMALS. McGraw-Hill Book Co. 1949.

Miner, R. W. FIELD BOOK OF SEASHORE LIFE. G. P. Putnam's Sons. 1950.

Morris, P. A. A FIELD GUIDE TO SHELLS OF THE ATLANTIC AND GULF COASTS. Houghton Mifflin Co. 1951.

Morris, P. A. A FIELD GUIDE TO SHELLS OF THE PACIFIC COAST. Houghton Mifflin Co. 1952.

Pope, C. H. TURTLES OF THE UNITED STATES AND CANADA. Alfred A. Knopf. 1939.

Ricketts, E. K. and J. Calvin. BETWEEN PACIFIC TIDES. Stanford University Press. 1948.

Russell, F. S. and C. M. Yonge. THE SEAS. Warne Co. London. 1936.

Wilson, D. P. LIFE OF THE SHORES AND SHALLOW SEA. Nicholson and Watson. London. 1935.

Yonge, C. M. THE SEA SHORE. Collins. London. 1949.

Zim, H. S. and L. Ingle. SEASHORES. Simon and Schuster. 1955.

INDEX

All references are to question numbers

Abalone, 341-355, 991, 1045
  black (*Haliotis cracherodii*), 342
  green (*Haliotis fulgens*), 342
  red (*Haliotis rufescens*), 342, 351-353
Acetabularia (mermaids' cup), 88, 115
Acmaea candeana (southern limpet), 311
  — *digitalis* (common Pacific limpet), 311
  — *paleacea* (chaffy limpet), 311
  — *pelta* (shield limpet), 311
  — *scabra* (ribbed limpet), 311
  — *testudinalis* (tortoise-shell limpet), 311
Acorn worm, 4, 902-905
Acropora (staghorn coral), 841
Adamsia tricolor (anemone), 633, 803
Aeolis papillosa (sea slug), 365, 366
Age of
  anemone, 795
  barnacles, 658
  clam, 394
  crab, 522
  limpet, 317
  sea hare, 361
  sea urchin, 206
  shells, 149
  shrimp, 596
  starfish, 189
Albatross, 1102
Alcyonium digitatum (finger coral), 841, 857
Aletes squamigerus (worm tube snail), 292
Algae
  blue-green, 110

  brown, 110, 111, 120
  green, 10, 110, 114, 115
  red, 110, 112
  stony, 852
Algin, 111, 125
American oyster; *see* Oyster, American
Amaroucium stellatum (sea pork), 919
Ammodytes (sand eel), 928, 950-952
Amphioxus (lancelet), 906-908
Amphipholis squamata (snake brittle star), 201
Amphitrite ornata (mud-tube worm), 750, 751, 752
Anemones, 5, 12, 13, 55, 59, 290, 633, 682, 758, 788-805, 844, 845, 847
  brooding (*Epiactis prolifera*), 797
  burrowing (*Harenactis attenuata*), 802, 859
  burrowing (*Cerianthus americanus*), 788, 802
  dahlia (*Tealia crassicornis*), 801
  great green (*Cribina xanthogrammica*), 794, 801
  plumose (*Metridium dianthus*), 788-797
  small green (*Bunodactis Elegantissima*), 797, 801
Angry crab; *see* Crabs, green shore
Anisodoris nobilis (sea slug), 365, 366
Anomia simplex (jingle shell), 444, 445
Anurida (marine insect), 683
Aphrodite aculeata (sea mouse), 734, 739-740
Arbacia punctulata (sea urchin), 216

A CATALOG OF SELECTED

DOVER BOOKS

IN ALL FIELDS OF INTEREST

A CATALOG OF SELECTED DOVER
BOOKS IN ALL FIELDS OF INTEREST

DRAWINGS OF REMBRANDT, edited by Seymour Slive. Updated Lippmann, Hofstede de Groot edition, with definitive scholarly apparatus. All portraits, biblical sketches, landscapes, nudes. Oriental figures, classical studies, together with selection of work by followers. 550 illustrations. Total of 630pp. 9⅛ × 12¼.
21485-0, 21486-9 Pa., Two-vol. set $25.00

GHOST AND HORROR STORIES OF AMBROSE BIERCE, Ambrose Bierce. 24 tales vividly imagined, strangely prophetic, and decades ahead of their time in technical skill: "The Damned Thing," "An Inhabitant of Carcosa," "The Eyes of the Panther," "Moxon's Master," and 20 more. 199pp. 5⅜ × 8½. 20767-6 Pa. $3.95

ETHICAL WRITINGS OF MAIMONIDES, Maimonides. Most significant ethical works of great medieval sage, newly translated for utmost precision, readability. Laws Concerning Character Traits, Eight Chapters, more. 192pp. 5⅜ × 8½.
24522-5 Pa. $4.50

THE EXPLORATION OF THE COLORADO RIVER AND ITS CANYONS, J. W. Powell. Full text of Powell's 1,000-mile expedition down the fabled Colorado in 1869. Superb account of terrain, geology, vegetation, Indians, famine, mutiny, treacherous rapids, mighty canyons, during exploration of last unknown part of continental U.S. 400pp. 5⅜ × 8½. 20094-9 Pa. $6.95

HISTORY OF PHILOSOPHY, Julián Marías. Clearest one-volume history on the market. Every major philosopher and dozens of others, to Existentialism and later. 505pp. 5⅜ × 8½. 21739-6 Pa. $8.50

ALL ABOUT LIGHTNING, Martin A. Uman. Highly readable non-technical survey of nature and causes of lightning, thunderstorms, ball lightning, St. Elmo's Fire, much more. Illustrated. 192pp. 5⅜ × 8½. 25237-X Pa. $5.95

SAILING ALONE AROUND THE WORLD, Captain Joshua Slocum. First man to sail around the world, alone, in small boat. One of great feats of seamanship told in delightful manner. 67 illustrations. 294pp. 5⅜ × 8½. 20326-3 Pa. $4.95

LETTERS AND NOTES ON THE MANNERS, CUSTOMS AND CONDITIONS OF THE NORTH AMERICAN INDIANS, George Catlin. Classic account of life among Plains Indians: ceremonies, hunt, warfare, etc. 312 plates. 572pp. of text. 6⅛ × 9¼. 22118-0, 22119-9 Pa. Two-vol. set $15.90

ALASKA: The Harriman Expedition, 1899, John Burroughs, John Muir, et al. Informative, engrossing accounts of two-month, 9,000-mile expedition. Native peoples, wildlife, forests, geography, salmon industry, glaciers, more. Profusely illustrated. 240 black-and-white line drawings. 124 black-and-white photographs. 3 maps. Index. 576pp. 5⅜ × 8½. 25109-8 Pa. $11.95

THE BOOK OF BEASTS: Being a Translation from a Latin Bestiary of the Twelfth Century, T. H. White. Wonderful catalog real and fanciful beasts: manticore, griffin, phoenix, amphivius, jaculus, many more. White's witty erudite commentary on scientific, historical aspects. Fascinating glimpse of medieval mind. Illustrated. 296pp. 5⅜ × 8¼. (Available in U.S. only) 24609-4 Pa. $5.95

FRANK LLOYD WRIGHT: ARCHITECTURE AND NATURE With 160 Illustrations, Donald Hoffmann. Profusely illustrated study of influence of nature—especially prairie—on Wright's designs for Fallingwater, Robie House, Guggenheim Museum, other masterpieces. 96pp. 9¼ × 10¾. 25098-9 Pa. $7.95

FRANK LLOYD WRIGHT'S FALLINGWATER, Donald Hoffmann. Wright's famous waterfall house: planning and construction of organic idea. History of site, owners, Wright's personal involvement. Photographs of various stages of building. Preface by Edgar Kaufmann, Jr. 100 illustrations. 112pp. 9¼ × 10. 23671-4 Pa. $7.95

YEARS WITH FRANK LLOYD WRIGHT: Apprentice to Genius, Edgar Tafel. Insightful memoir by a former apprentice presents a revealing portrait of Wright the man, the inspired teacher, the greatest American architect. 372 black-and-white illustrations. Preface. Index. vi + 228pp. 8¼ × 11. 24801-1 Pa. $9.95

THE STORY OF KING ARTHUR AND HIS KNIGHTS, Howard Pyle. Enchanting version of King Arthur fable has delighted generations with imaginative narratives of exciting adventures and unforgettable illustrations by the author. 41 illustrations. xviii + 313pp. 6⅛ × 9¼. 21445-1 Pa. $5.95

THE GODS OF THE EGYPTIANS, E. A. Wallis Budge. Thorough coverage of numerous gods of ancient Egypt by foremost Egyptologist. Information on evolution of cults, rites and gods; the cult of Osiris; the Book of the Dead and its rites; the sacred animals and birds; Heaven and Hell; and more. 956pp. 6⅛ × 9¼. 22055-9, 22056-7 Pa., Two-vol. set $21.90

A THEOLOGICO-POLITICAL TREATISE, Benedict Spinoza. Also contains unfinished Political Treatise. Great classic on religious liberty, theory of government on common consent. R. Elwes translation. Total of 421pp. 5⅜ × 8½. 20249-6 Pa. $6.95

INCIDENTS OF TRAVEL IN CENTRAL AMERICA, CHIAPAS, AND YUCATAN, John L. Stephens. Almost single-handed discovery of Maya culture; exploration of ruined cities, monuments, temples; customs of Indians. 115 drawings. 892pp. 5⅜ × 8½. 22404-X, 22405-8 Pa., Two-vol. set $15.90

LOS CAPRICHOS, Francisco Goya. 80 plates of wild, grotesque monsters and caricatures. Prado manuscript included. 183pp. 6⅞ × 9⅜. 22384-1 Pa. $4.95

AUTOBIOGRAPHY: The Story of My Experiments with Truth, Mohandas K. Gandhi. Not hagiography, but Gandhi in his own words. Boyhood, legal studies, purification, the growth of the Satyagraha (nonviolent protest) movement. Critical, inspiring work of the man who freed India. 480pp. 5⅜ × 8½. (Available in U.S. only) 24593-4 Pa. $6.95

ILLUSTRATED DICTIONARY OF HISTORIC ARCHITECTURE, edited by Cyril M. Harris. Extraordinary compendium of clear, concise definitions for over 5,000 important architectural terms complemented by over 2,000 line drawings. Covers full spectrum of architecture from ancient ruins to 20th-century Modernism. Preface. 592pp. 7½ × 9⅝. 24444-X Pa. $14.95

THE NIGHT BEFORE CHRISTMAS, Clement Moore. Full text, and woodcuts from original 1848 book. Also critical, historical material. 19 illustrations. 40pp. 4⅝ × 6. 22797-9 Pa. $2.50

THE LESSON OF JAPANESE ARCHITECTURE: 165 Photographs, Jiro Harada. Memorable gallery of 165 photographs taken in the 1930's of exquisite Japanese homes of the well-to-do and historic buildings. 13 line diagrams. 192pp. 8⅜ × 11¼. 24778-3 Pa. $8.95

THE AUTOBIOGRAPHY OF CHARLES DARWIN AND SELECTED LETTERS, edited by Francis Darwin. The fascinating life of eccentric genius composed of an intimate memoir by Darwin (intended for his children); commentary by his son, Francis; hundreds of fragments from notebooks, journals, papers; and letters to and from Lyell, Hooker, Huxley, Wallace and Henslow. xi + 365pp. 5⅜ × 8. 20479-0 Pa. $5.95

WONDERS OF THE SKY: Observing Rainbows, Comets, Eclipses, the Stars and Other Phenomena, Fred Schaaf. Charming, easy-to-read poetic guide to all manner of celestial events visible to the naked eye. Mock suns, glories, Belt of Venus, more. Illustrated. 299pp. 5¼ × 8¼. 24402-4 Pa. $7.95

BURNHAM'S CELESTIAL HANDBOOK, Robert Burnham, Jr. Thorough guide to the stars beyond our solar system. Exhaustive treatment. Alphabetical by constellation: Andromeda to Cetus in Vol. 1; Chamaeleon to Orion in Vol. 2; and Pavo to Vulpecula in Vol. 3. Hundreds of illustrations. Index in Vol. 3. 2,000pp. 6⅛ × 9¼. 23567-X, 23568-8, 23673-0 Pa., Three-vol. set $37.85

STAR NAMES: Their Lore and Meaning, Richard Hinckley Allen. Fascinating history of names various cultures have given to constellations and literary and folkloristic uses that have been made of stars. Indexes to subjects. Arabic and Greek names. Biblical references. Bibliography. 563pp. 5⅜ × 8½. 21079-0 Pa. $7.95

THIRTY YEARS THAT SHOOK PHYSICS: The Story of Quantum Theory, George Gamow. Lucid, accessible introduction to influential theory of energy and matter. Careful explanations of Dirac's anti-particles, Bohr's model of the atom, much more. 12 plates. Numerous drawings. 240pp. 5⅜ × 8½. 24895-X Pa. $4.95

CHINESE DOMESTIC FURNITURE IN PHOTOGRAPHS AND MEASURED DRAWINGS, Gustav Ecke. A rare volume, now affordably priced for antique collectors, furniture buffs and art historians. Detailed review of styles ranging from early Shang to late Ming. Unabridged republication. 161 black-and-white drawings, photos. Total of 224pp. 8⅜ × 11¼. (Available in U.S. only) 25171-3 Pa. $12.95

VINCENT VAN GOGH: A Biography, Julius Meier-Graefe. Dynamic, penetrating study of artist's life, relationship with brother, Theo, painting techniques, travels, more. Readable, engrossing. 160pp. 5⅜ × 8½. (Available in U.S. only) 25253-1 Pa. $3.95

HOW TO WRITE, Gertrude Stein. Gertrude Stein claimed anyone could understand her unconventional writing—here are clues to help. Fascinating improvisations, language experiments, explanations illuminate Stein's craft and the art of writing. Total of 414pp. 4⅝ × 6⅝. 23144-5 Pa. $5.95

ADVENTURES AT SEA IN THE GREAT AGE OF SAIL: Five Firsthand Narratives, edited by Elliot Snow. Rare true accounts of exploration, whaling, shipwreck, fierce natives, trade, shipboard life, more. 33 illustrations. Introduction. 353pp. 5⅜ × 8½. 25177-2 Pa. $7.95

THE HERBAL OR GENERAL HISTORY OF PLANTS, John Gerard. Classic descriptions of about 2,850 plants—with over 2,700 illustrations—includes Latin and English names, physical descriptions, varieties, time and place of growth, more. 2,706 illustrations. xlv + 1,678pp. 8½ × 12¼. 23147-X Cloth. $75.00

DOROTHY AND THE WIZARD IN OZ, L. Frank Baum. Dorothy and the Wizard visit the center of the Earth, where people are vegetables, glass houses grow and Oz characters reappear. Classic sequel to *Wizard of Oz*. 256pp. 5⅜ × 8. 24714-7 Pa. $4.95

SONGS OF EXPERIENCE: Facsimile Reproduction with 26 Plates in Full Color, William Blake. This facsimile of Blake's original "Illuminated Book" reproduces 26 full-color plates from a rare 1826 edition. Includes "The Tyger," "London," "Holy Thursday," and other immortal poems. 26 color plates. Printed text of poems. 48pp. 5¼ × 7. 24636-1 Pa. $3.50

SONGS OF INNOCENCE, William Blake. The first and most popular of Blake's famous "Illuminated Books," in a facsimile edition reproducing all 31 brightly colored plates. Additional printed text of each poem. 64pp. 5¼ × 7. 22764-2 Pa. $3.50

PRECIOUS STONES, Max Bauer. Classic, thorough study of diamonds, rubies, emeralds, garnets, etc.: physical character, occurrence, properties, use, similar topics. 20 plates, 8 in color. 94 figures. 659pp. 6⅛ × 9¼. 21910-0, 21911-9 Pa., Two-vol. set $15.90

ENCYCLOPEDIA OF VICTORIAN NEEDLEWORK, S. F. A. Caulfeild and Blanche Saward. Full, precise descriptions of stitches, techniques for dozens of needlecrafts—most exhaustive reference of its kind. Over 800 figures. Total of 679pp. 8⅛ × 11. Two volumes. Vol. 1 22800-2 Pa. $11.95 Vol. 2 22801-0 Pa. $11.95

THE MARVELOUS LAND OF OZ, L. Frank Baum. Second Oz book, the Scarecrow and Tin Woodman are back with hero named Tip, Oz magic. 136 illustrations. 287pp. 5⅜ × 8½. 20692-0 Pa. $5.95

WILD FOWL DECOYS, Joel Barber. Basic book on the subject, by foremost authority and collector. Reveals history of decoy making and rigging, place in American culture, different kinds of decoys, how to make them, and how to use them. 140 plates. 156pp. 7⅞ × 10¾. 20011-6 Pa. $8.95

HISTORY OF LACE, Mrs. Bury Palliser. Definitive, profusely illustrated chronicle of lace from earliest times to late 19th century. Laces of Italy, Greece, England, France, Belgium, etc. Landmark of needlework scholarship. 266 illustrations. 672pp. 6⅛ × 9¼. 24742-2 Pa. $14.95

ILLUSTRATED GUIDE TO SHAKER FURNITURE, Robert Meader. All furniture and appurtenances, with much on unknown local styles. 235 photos. 146pp. 9 × 12. 22819-3 Pa. $7.95

WHALE SHIPS AND WHALING: A Pictorial Survey, George Francis Dow. Over 200 vintage engravings, drawings, photographs of barks, brigs, cutters, other vessels. Also harpoons, lances, whaling guns, many other artifacts. Comprehensive text by foremost authority. 207 black-and-white illustrations. 288pp. 6 × 9. 24808-9 Pa. $8.95

THE BERTRAMS, Anthony Trollope. Powerful portrayal of blind self-will and thwarted ambition includes one of Trollope's most heartrending love stories. 497pp. 5⅜ × 8½. 25119-5 Pa. $8.95

ADVENTURES WITH A HAND LENS, Richard Headstrom. Clearly written guide to observing and studying flowers and grasses, fish scales, moth and insect wings, egg cases, buds, feathers, seeds, leaf scars, moss, molds, ferns, common crystals, etc.—all with an ordinary, inexpensive magnifying glass. 209 exact line drawings aid in your discoveries. 220pp. 5⅜ × 8½. 23330-8 Pa. $4.50

RODIN ON ART AND ARTISTS, Auguste Rodin. Great sculptor's candid, wide-ranging comments on meaning of art; great artists; relation of sculpture to poetry, painting, music; philosophy of life, more. 76 superb black-and-white illustrations of Rodin's sculpture, drawings and prints. 119pp. 8⅜ × 11¼. 24487-3 Pa. $6.95

FIFTY CLASSIC FRENCH FILMS, 1912–1982: A Pictorial Record, Anthony Slide. Memorable stills from Grand Illusion, Beauty and the Beast, Hiroshima, Mon Amour, many more. Credits, plot synopses, reviews, etc. 160pp. 8¼ × 11. 25256-6 Pa. $11.95

THE PRINCIPLES OF PSYCHOLOGY, William James. Famous long course complete, unabridged. Stream of thought, time perception, memory, experimental methods; great work decades ahead of its time. 94 figures. 1,391pp. 5⅜ × 8½. 20381-6, 20382-4 Pa., Two-vol. set $19.90

BODIES IN A BOOKSHOP, R. T. Campbell. Challenging mystery of blackmail and murder with ingenious plot and superbly drawn characters. In the best tradition of British suspense fiction. 192pp. 5⅜ × 8½. 24720-1 Pa. $3.95

CALLAS: PORTRAIT OF A PRIMA DONNA, George Jellinek. Renowned commentator on the musical scene chronicles incredible career and life of the most controversial, fascinating, influential operatic personality of our time. 64 black-and-white photographs. 416pp. 5⅜ × 8¼. 25047-4 Pa. $7.95

GEOMETRY, RELATIVITY AND THE FOURTH DIMENSION, Rudolph Rucker. Exposition of fourth dimension, concepts of relativity as Flatland characters continue adventures. Popular, easily followed yet accurate, profound. 141 illustrations. 133pp. 5⅜ × 8½. 23400-2 Pa. $3.50

HOUSEHOLD STORIES BY THE BROTHERS GRIMM, with pictures by Walter Crane. 53 classic stories—Rumpelstiltskin, Rapunzel, Hansel and Gretel, the Fisherman and his Wife, Snow White, Tom Thumb, Sleeping Beauty, Cinderella, and so much more—lavishly illustrated with original 19th century drawings. 114 illustrations. x + 269pp. 5⅜ × 8½. 21080-4 Pa. $4.50

SUNDIALS, Albert Waugh. Far and away the best, most thorough coverage of ideas, mathematics concerned, types, construction, adjusting anywhere. Over 100 illustrations. 230pp. 5⅜ × 8½. 22947-5 Pa. $4.50

PICTURE HISTORY OF THE NORMANDIE: With 190 Illustrations, Frank O. Braynard. Full story of legendary French ocean liner: Art Deco interiors, design innovations, furnishings, celebrities, maiden voyage, tragic fire, much more. Extensive text. 144pp. 8⅞ × 11¾. 25257-4 Pa. $9.95

THE FIRST AMERICAN COOKBOOK: A Facsimile of "American Cookery," 1796, Amelia Simmons. Facsimile of the first American-written cookbook published in the United States contains authentic recipes for colonial favorites—pumpkin pudding, winter squash pudding, spruce beer, Indian slapjacks, and more. Introductory Essay and Glossary of colonial cooking terms. 80pp. 5⅜ × 8½. 24710-4 Pa. $3.50

101 PUZZLES IN THOUGHT AND LOGIC, C. R. Wylie, Jr. Solve murders and robberies, find out which fishermen are liars, how a blind man could possibly identify a color—purely by your own reasoning! 107pp. 5⅜ × 8½. 20367-0 Pa. $2.50

THE BOOK OF WORLD-FAMOUS MUSIC—CLASSICAL, POPULAR AND FOLK, James J. Fuld. Revised and enlarged republication of landmark work in musico-bibliography. Full information about nearly 1,000 songs and compositions including first lines of music and lyrics. New supplement. Index. 800pp. 5⅜ × 8¼. 24857-7 Pa. $14.95

ANTHROPOLOGY AND MODERN LIFE, Franz Boas. Great anthropologist's classic treatise on race and culture. Introduction by Ruth Bunzel. Only inexpensive paperback edition. 255pp. 5⅜ × 8½. 25245-0 Pa. $5.95

THE TALE OF PETER RABBIT, Beatrix Potter. The inimitable Peter's terrifying adventure in Mr. McGregor's garden, with all 27 wonderful, full-color Potter illustrations. 55pp. 4¼ × 5½. (Available in U.S. only) 22827-4 Pa. $1.75

THREE PROPHETIC SCIENCE FICTION NOVELS, H. G. Wells. *When the Sleeper Wakes, A Story of the Days to Come* and *The Time Machine* (full version). 335pp. 5⅜ × 8½. (Available in U.S. only) 20605-X Pa. $5.95

APICIUS COOKERY AND DINING IN IMPERIAL ROME, edited and translated by Joseph Dommers Vehling. Oldest known cookbook in existence offers readers a clear picture of what foods Romans ate, how they prepared them, etc. 49 illustrations. 301pp. 6⅛ × 9¼. 23563-7 Pa. $6.50

SHAKESPEARE LEXICON AND QUOTATION DICTIONARY, Alexander Schmidt. Full definitions, locations, shades of meaning of every word in plays and poems. More than 50,000 exact quotations. 1,485pp. 6½ × 9¼. 22726-X, 22727-8 Pa., Two-vol. set $27.90

THE WORLD'S GREAT SPEECHES, edited by Lewis Copeland and Lawrence W. Lamm. Vast collection of 278 speeches from Greeks to 1970. Powerful and effective models; unique look at history. 842pp. 5⅜ × 8½. 20468-5 Pa. $11.95

THE BLUE FAIRY BOOK, Andrew Lang. The first, most famous collection, with many familiar tales: Little Red Riding Hood, Aladdin and the Wonderful Lamp, Puss in Boots, Sleeping Beauty, Hansel and Gretel, Rumpelstiltskin; 37 in all. 138 illustrations. 390pp. 5⅜ × 8½. 21437-0 Pa. $5.95

THE STORY OF THE CHAMPIONS OF THE ROUND TABLE, Howard Pyle. Sir Launcelot, Sir Tristram and Sir Percival in spirited adventures of love and triumph retold in Pyle's inimitable style. 50 drawings, 31 full-page. xviii + 329pp. 6½ × 9¼. 21883-X Pa. $6.95

AUDUBON AND HIS JOURNALS, Maria Audubon. Unmatched two-volume portrait of the great artist, naturalist and author contains his journals, an excellent biography by his granddaughter, expert annotations by the noted ornithologist, Dr. Elliott Coues, and 37 superb illustrations. Total of 1,200pp. 5⅜ × 8.
Vol. I 25143-8 Pa. $8.95
Vol. II 25144-6 Pa. $8.95

GREAT DINOSAUR HUNTERS AND THEIR DISCOVERIES, Edwin H. Colbert. Fascinating, lavishly illustrated chronicle of dinosaur research, 1820's to 1960. Achievements of Cope, Marsh, Brown, Buckland, Mantell, Huxley, many others. 384pp. 5¼ × 8¼. 24701-5 Pa. $6.95

THE TASTEMAKERS, Russell Lynes. Informal, illustrated social history of American taste 1850's–1950's. First popularized categories Highbrow, Lowbrow, Middlebrow. 129 illustrations. New (1979) afterword. 384pp. 6 × 9.
23993-4 Pa. $6.95

DOUBLE CROSS PURPOSES, Ronald A. Knox. A treasure hunt in the Scottish Highlands, an old map, unidentified corpse, surprise discoveries keep reader guessing in this cleverly intricate tale of financial skullduggery. 2 black-and-white maps. 320pp. 5⅜ × 8½. (Available in U.S. only) 25032-6 Pa. $5.95

AUTHENTIC VICTORIAN DECORATION AND ORNAMENTATION IN FULL COLOR: 46 Plates from "Studies in Design," Christopher Dresser. Superb full-color lithographs reproduced from rare original portfolio of a major Victorian designer. 48pp. 9¼ × 12¼. 25083-0 Pa. $7.95

PRIMITIVE ART, Franz Boas. Remains the best text ever prepared on subject, thoroughly discussing Indian, African, Asian, Australian, and, especially, Northern American primitive art. Over 950 illustrations show ceramics, masks, totem poles, weapons, textiles, paintings, much more. 376pp. 5⅜ × 8. 20025-6 Pa. $6.95

SIDELIGHTS ON RELATIVITY, Albert Einstein. Unabridged republication of two lectures delivered by the great physicist in 1920–21. *Ether and Relativity* and *Geometry and Experience*. Elegant ideas in non-mathematical form, accessible to intelligent layman. vi + 56pp. 5⅜ × 8½. 24511-X Pa. $2.95

THE WIT AND HUMOR OF OSCAR WILDE, edited by Alvin Redman. More than 1,000 ripostes, paradoxes, wisecracks: Work is the curse of the drinking classes, I can resist everything except temptation, etc. 258pp. 5⅜ × 8½. 20602-5 Pa. $4.50

ADVENTURES WITH A MICROSCOPE, Richard Headstrom. 59 adventures with clothing fibers, protozoa, ferns and lichens, roots and leaves, much more. 142 illustrations. 232pp. 5⅜ × 8½. 23471-1 Pa. $3.95

CATALOG OF DOVER BOOKS

PLANTS OF THE BIBLE, Harold N. Moldenke and Alma L. Moldenke. Standard reference to all 230 plants mentioned in Scriptures. Latin name, biblical reference, uses, modern identity, much more. Unsurpassed encyclopedic resource for scholars, botanists, nature lovers, students of Bible. Bibliography. Indexes. 123 black-and-white illustrations. 384pp. 6 × 9.   25069-5 Pa. $8.95

FAMOUS AMERICAN WOMEN: A Biographical Dictionary from Colonial Times to the Present, Robert McHenry, ed. From Pocahontas to Rosa Parks, 1,035 distinguished American women documented in separate biographical entries. Accurate, up-to-date data, numerous categories, spans 400 years. Indices. 493pp. 6½ × 9¼.   24523-3 Pa. $9.95

THE FABULOUS INTERIORS OF THE GREAT OCEAN LINERS IN HISTORIC PHOTOGRAPHS, William H. Miller, Jr. Some 200 superb photographs capture exquisite interiors of world's great "floating palaces"—1890's to 1980's: *Titanic, Ile de France, Queen Elizabeth, United States, Europa,* more. Approx. 200 black-and-white photographs. Captions. Text. Introduction. 160pp. 8⅜ × 11¼.
24756-2 Pa. $9.95

THE GREAT LUXURY LINERS, 1927–1954: A Photographic Record, William H. Miller, Jr. Nostalgic tribute to heyday of ocean liners. 186 photos of Ile de France, Normandie, Leviathan, Queen Elizabeth, United States, many others. Interior and exterior views. Introduction. Captions. 160pp. 9 × 12.
24056-8 Pa. $9.95

A NATURAL HISTORY OF THE DUCKS, John Charles Phillips. Great landmark of ornithology offers complete detailed coverage of nearly 200 species and subspecies of ducks: gadwall, sheldrake, merganser, pintail, many more. 74 full-color plates, 102 black-and-white. Bibliography. Total of 1,920pp. 8⅜ × 11¼.
25141-1, 25142-X Cloth. Two-vol. set $100.00

THE SEAWEED HANDBOOK: An Illustrated Guide to Seaweeds from North Carolina to Canada, Thomas F. Lee. Concise reference covers 78 species. Scientific and common names, habitat, distribution, more. Finding keys for easy identification. 224pp. 5⅜ × 8½.   25215-9 Pa. $5.95

THE TEN BOOKS OF ARCHITECTURE: The 1755 Leoni Edition, Leon Battista Alberti. Rare classic helped introduce the glories of ancient architecture to the Renaissance. 68 black-and-white plates. 336pp. 8⅜ × 11¼.   25239-6 Pa. $14.95

MISS MACKENZIE, Anthony Trollope. Minor masterpieces by Victorian master unmasks many truths about life in 19th-century England. First inexpensive edition in years. 392pp. 5⅜ × 8½.   25201-9 Pa. $7.95

THE RIME OF THE ANCIENT MARINER, Gustave Doré, Samuel Taylor Coleridge. Dramatic engravings considered by many to be his greatest work. The terrifying space of the open sea, the storms and whirlpools of an unknown ocean, the ice of Antarctica, more—all rendered in a powerful, chilling manner. Full text. 38 plates. 77pp. 9¼ × 12.   22305-1 Pa. $4.95

THE EXPEDITIONS OF ZEBULON MONTGOMERY PIKE, Zebulon Montgomery Pike. Fascinating first-hand accounts (1805–6) of exploration of Mississippi River, Indian wars, capture by Spanish dragoons, much more. 1,088pp. 5⅜ × 8½.   25254-X, 25255-8 Pa. Two-vol. set $23.90

A CONCISE HISTORY OF PHOTOGRAPHY: Third Revised Edition, Helmut Gernsheim. Best one-volume history—camera obscura, photochemistry, daguerreotypes, evolution of cameras, film, more. Also artistic aspects—landscape, portraits, fine art, etc. 281 black-and-white photographs. 26 in color. 176pp. 8⅜ × 11¼. 25128-4 Pa. $12.95

THE DORÉ BIBLE ILLUSTRATIONS, Gustave Doré. 241 detailed plates from the Bible: the Creation scenes, Adam and Eve, Flood, Babylon, battle sequences, life of Jesus, etc. Each plate is accompanied by the verses from the King James version of the Bible. 241pp. 9 × 12. 23004-X Pa. $8.95

HUGGER-MUGGER IN THE LOUVRE, Elliot Paul. Second Homer Evans mystery-comedy. Theft at the Louvre involves sleuth in hilarious, madcap caper. "A knockout."—Books. 336pp. 5⅜ × 8½. 25185-3 Pa. $5.95

FLATLAND, E. A. Abbott. Intriguing and enormously popular science-fiction classic explores the complexities of trying to survive as a two-dimensional being in a three-dimensional world. Amusingly illustrated by the author. 16 illustrations. 103pp. 5⅜ × 8½. 20001-9 Pa. $2.25

THE HISTORY OF THE LEWIS AND CLARK EXPEDITION, Meriwether Lewis and William Clark, edited by Elliott Coues. Classic edition of Lewis and Clark's day-by-day journals that later became the basis for U.S. claims to Oregon and the West. Accurate and invaluable geographical, botanical, biological, meteorological and anthropological material. Total of 1,508pp. 5⅜ × 8½.
21268-8, 21269-6, 21270-X Pa. Three-vol. set $25.50

LANGUAGE, TRUTH AND LOGIC, Alfred J. Ayer. Famous, clear introduction to Vienna, Cambridge schools of Logical Positivism. Role of philosophy, elimination of metaphysics, nature of analysis, etc. 160pp. 5⅜ × 8½. (Available in U.S. and Canada only) 20010-8 Pa. $2.95

MATHEMATICS FOR THE NONMATHEMATICIAN, Morris Kline. Detailed, college-level treatment of mathematics in cultural and historical context, with numerous exercises. For liberal arts students. Preface. Recommended Reading Lists. Tables. Index. Numerous black-and-white figures. xvi + 641pp. 5⅜ × 8½.
24823-2 Pa. $11.95

28 SCIENCE FICTION STORIES, H. G. Wells. Novels, *Star Begotten* and *Men Like Gods,* plus 26 short stories: "Empire of the Ants," "A Story of the Stone Age," "The Stolen Bacillus," "In the Abyss," etc. 915pp. 5⅜ × 8½. (Available in U.S. only)
20265-8 Cloth. $10.95

HANDBOOK OF PICTORIAL SYMBOLS, Rudolph Modley. 3,250 signs and symbols, many systems in full; official or heavy commercial use. Arranged by subject. Most in Pictorial Archive series. 143pp. 8⅛ × 11. 23357-X Pa. $5.95

INCIDENTS OF TRAVEL IN YUCATAN, John L. Stephens. Classic (1843) exploration of jungles of Yucatan, looking for evidences of Maya civilization. Travel adventures, Mexican and Indian culture, etc. Total of 669pp. 5⅜ × 8½.
20926-1, 20927-X Pa., Two-vol. set $9.90

CATALOG OF DOVER BOOKS

DEGAS: An Intimate Portrait, Ambroise Vollard. Charming, anecdotal memoir by famous art dealer of one of the greatest 19th-century French painters. 14 black-and-white illustrations. Introduction by Harold L. Van Doren. 96pp. 5⅜ × 8½.
25131-4 Pa. $3.95

PERSONAL NARRATIVE OF A PILGRIMAGE TO ALMANDINAH AND MECCAH, Richard Burton. Great travel classic by remarkably colorful personality. Burton, disguised as a Moroccan, visited sacred shrines of Islam, narrowly escaping death. 47 illustrations. 959pp. 5⅜ × 8½. 21217-3, 21218-1 Pa., Two-vol. set $17.90

PHRASE AND WORD ORIGINS, A. H. Holt. Entertaining, reliable, modern study of more than 1,200 colorful words, phrases, origins and histories. Much unexpected information. 254pp. 5⅜ × 8½. 20758-7 Pa. $5.95

THE RED THUMB MARK, R. Austin Freeman. In this first Dr. Thorndyke case, the great scientific detective draws fascinating conclusions from the nature of a single fingerprint. Exciting story, authentic science. 320pp. 5⅜ × 8½. (Available in U.S. only) 25210-8 Pa. $5.95

AN EGYPTIAN HIEROGLYPHIC DICTIONARY, E. A. Wallis Budge. Monumental work containing about 25,000 words or terms that occur in texts ranging from 3000 B.C. to 600 A.D. Each entry consists of a transliteration of the word, the word in hieroglyphs, and the meaning in English. 1,314pp. 6⅜ × 10.
23615-3, 23616-1 Pa., Two-vol. set $27.90

THE COMPLEAT STRATEGYST: Being a Primer on the Theory of Games of Strategy, J. D. Williams. Highly entertaining classic describes, with many illustrated examples, how to select best strategies in conflict situations. Prefaces. Appendices. xvi + 268pp. 5⅜ × 8½. 25101-2 Pa. $5.95

THE ROAD TO OZ, L. Frank Baum. Dorothy meets the Shaggy Man, little Button-Bright and the Rainbow's beautiful daughter in this delightful trip to the magical Land of Oz. 272pp. 5⅜ × 8. 25208-6 Pa. $4.95

POINT AND LINE TO PLANE, Wassily Kandinsky. Seminal exposition of role of point, line, other elements in non-objective painting. Essential to understanding 20th-century art. 127 illustrations. 192pp. 6½ × 9¼. 23808-3 Pa. $4.50

LADY ANNA, Anthony Trollope. Moving chronicle of Countess Lovel's bitter struggle to win for herself and daughter Anna their rightful rank and fortune—perhaps at cost of sanity itself. 384pp. 5⅜ × 8½. 24669-8 Pa. $6.95

EGYPTIAN MAGIC, E. A. Wallis Budge. Sums up all that is known about magic in Ancient Egypt: the role of magic in controlling the gods, powerful amulets that warded off evil spirits, scarabs of immortality, use of wax images, formulas and spells, the secret name, much more. 253pp. 5⅜ × 8½. 22681-6 Pa. $4.50

THE DANCE OF SIVA, Ananda Coomaraswamy. Preeminent authority unfolds the vast metaphysic of India: the revelation of her art, conception of the universe, social organization, etc. 27 reproductions of art masterpieces. 192pp. 5⅜ × 8½.
24817-8 Pa. $5.95

CATALOG OF DOVER BOOKS

CHRISTMAS CUSTOMS AND TRADITIONS, Clement A. Miles. Origin, evolution, significance of religious, secular practices. Caroling, gifts, yule logs, much more. Full, scholarly yet fascinating; non-sectarian. 400pp. 5⅜ × 8½.
23354-5 Pa. $6.50

THE HUMAN FIGURE IN MOTION, Eadweard Muybridge. More than 4,500 stopped-action photos, in action series, showing undraped men, women, children jumping, lying down, throwing, sitting, wrestling, carrying, etc. 390pp. 7⅞ × 10⅝.
20204-6 Cloth. $19.95

THE MAN WHO WAS THURSDAY, Gilbert Keith Chesterton. Witty, fast-paced novel about a club of anarchists in turn-of-the-century London. Brilliant social, religious, philosophical speculations. 128pp. 5⅜ × 8½.
25121-7 Pa. $3.95

A CEZANNE SKETCHBOOK: Figures, Portraits, Landscapes and Still Lifes, Paul Cezanne. Great artist experiments with tonal effects, light, mass, other qualities in over 100 drawings. A revealing view of developing master painter, precursor of Cubism. 102 black-and-white illustrations. 144pp. 8¾ × 6⅝.
24790-2 Pa. $5.95

AN ENCYCLOPEDIA OF BATTLES: Accounts of Over 1,560 Battles from 1479 B.C. to the Present, David Eggenberger. Presents essential details of every major battle in recorded history, from the first battle of Megiddo in 1479 B.C. to Grenada in 1984. List of Battle Maps. New Appendix covering the years 1967–1984. Index. 99 illustrations. 544pp. 6½ × 9¼.
24913-1 Pa. $14.95

AN ETYMOLOGICAL DICTIONARY OF MODERN ENGLISH, Ernest Weekley. Richest, fullest work, by foremost British lexicographer. Detailed word histories. Inexhaustible. Total of 856pp. 6½ × 9¼.
21873-2, 21874-0 Pa., Two-vol. set $17.00

WEBSTER'S AMERICAN MILITARY BIOGRAPHIES, edited by Robert McHenry. Over 1,000 figures who shaped 3 centuries of American military history. Detailed biographies of Nathan Hale, Douglas MacArthur, Mary Hallaren, others. Chronologies of engagements, more. Introduction. Addenda. 1,033 entries in alphabetical order. xi + 548pp. 6½ × 9¼. (Available in U.S. only)
24758-9 Pa. $11.95

LIFE IN ANCIENT EGYPT, Adolf Erman. Detailed older account, with much not in more recent books: domestic life, religion, magic, medicine, commerce, and whatever else needed for complete picture. Many illustrations. 597pp. 5⅜ × 8½.
22632-8 Pa. $8.95

HISTORIC COSTUME IN PICTURES, Braun & Schneider. Over 1,450 costumed figures shown, covering a wide variety of peoples: kings, emperors, nobles, priests, servants, soldiers, scholars, townsfolk, peasants, merchants, courtiers, cavaliers, and more. 256pp. 8⅜ × 11¼.
23150-X Pa. $7.95

THE NOTEBOOKS OF LEONARDO DA VINCI, edited by J. P. Richter. Extracts from manuscripts reveal great genius; on painting, sculpture, anatomy, sciences, geography, etc. Both Italian and English. 186 ms. pages reproduced, plus 500 additional drawings, including studies for *Last Supper*, *Sforza* monument, etc. 860pp. 7⅞ × 10¾. (Available in U.S. only) 22572-0, 22573-9 Pa., Two-vol. set $25.90

CATALOG OF DOVER BOOKS

THE ART NOUVEAU STYLE BOOK OF ALPHONSE MUCHA: All 72 Plates from "Documents Decoratifs" in Original Color, Alphonse Mucha. Rare copyright-free design portfolio by high priest of Art Nouveau. Jewelry, wallpaper, stained glass, furniture, figure studies, plant and animal motifs, etc. Only complete one-volume edition. 80pp. 9⅜ × 12¼. 24044-4 Pa. $8.95

ANIMALS: 1,419 COPYRIGHT-FREE ILLUSTRATIONS OF MAMMALS, BIRDS, FISH, INSECTS, ETC., edited by Jim Harter. Clear wood engravings present, in extremely lifelike poses, over 1,000 species of animals. One of the most extensive pictorial sourcebooks of its kind. Captions. Index. 284pp. 9 × 12. 23766-4 Pa. $9.95

OBELISTS FLY HIGH, C. Daly King. Masterpiece of American detective fiction, long out of print, involves murder on a 1935 transcontinental flight—"a very thrilling story"—NY Times. Unabridged and unaltered republication of the edition published by William Collins Sons & Co. Ltd., London, 1935. 288pp. 5⅜ × 8½. (Available in U.S. only) 25036-9 Pa. $4.95

VICTORIAN AND EDWARDIAN FASHION: A Photographic Survey, Alison Gernsheim. First fashion history completely illustrated by contemporary photographs. Full text plus 235 photos, 1840-1914, in which many celebrities appear. 240pp. 6½ × 9¼. 24205-6 Pa. $6.00

THE ART OF THE FRENCH ILLUSTRATED BOOK, 1700-1914, Gordon N. Ray. Over 630 superb book illustrations by Fragonard, Delacroix, Daumier, Doré, Grandville, Manet, Mucha, Steinlen, Toulouse-Lautrec and many others. Preface. Introduction. 633 halftones. Indices of artists, authors & titles, binders and provenances. Appendices. Bibliography. 608pp. 8⅜ × 11¼. 25086-5 Pa. $24.95

THE WONDERFUL WIZARD OF OZ, L. Frank Baum. Facsimile in full color of America's finest children's classic. 143 illustrations by W. W. Denslow. 267pp. 5⅜ × 8½. 20691-2 Pa. $5.95

FRONTIERS OF MODERN PHYSICS: New Perspectives on Cosmology, Relativity, Black Holes and Extraterrestrial Intelligence, Tony Rothman, et al. For the intelligent layman. Subjects include: cosmological models of the universe; black holes; the neutrino; the search for extraterrestrial intelligence. Introduction. 46 black-and-white illustrations. 192pp. 5⅜ × 8½. 24587-X Pa. $6.95

THE FRIENDLY STARS, Martha Evans Martin & Donald Howard Menzel. Classic text marshalls the stars together in an engaging, non-technical survey, presenting them as sources of beauty in night sky. 23 illustrations. Foreword. 2 star charts. Index. 147pp. 5⅜ × 8½. 21099-5 Pa. $3.50

FADS AND FALLACIES IN THE NAME OF SCIENCE, Martin Gardner. Fair, witty appraisal of cranks, quacks, and quackeries of science and pseudoscience: hollow earth, Velikovsky, orgone energy, Dianetics, flying saucers, Bridey Murphy, food and medical fads, etc. Revised, expanded In the Name of Science. "A very able and even-tempered presentation."—The New Yorker. 363pp. 5⅜ × 8. 20394-8 Pa. $6.50

ANCIENT EGYPT: ITS CULTURE AND HISTORY, J. E Manchip White. From pre-dynastics through Ptolemies: society, history, political structure, religion, daily life, literature, cultural heritage. 48 plates. 217pp. 5⅜ × 8½. 22548-8 Pa. $4.95

CATALOG OF DOVER BOOKS

SIR HARRY HOTSPUR OF HUMBLETHWAITE, Anthony Trollope. Incisive, unconventional psychological study of a conflict between a wealthy baronet, his idealistic daughter, and their scapegrace cousin. The 1870 novel in its first inexpensive edition in years. 250pp. 5⅜ × 8½. 24953-0 Pa. $5.95

LASERS AND HOLOGRAPHY, Winston E. Kock. Sound introduction to burgeoning field, expanded (1981) for second edition. Wave patterns, coherence, lasers, diffraction, zone plates, properties of holograms, recent advances. 84 illustrations. 160pp. 5⅜ × 8¼. (Except in United Kingdom) 24041-X Pa. $3.50

INTRODUCTION TO ARTIFICIAL INTELLIGENCE: SECOND, EN-LARGED EDITION, Philip C. Jackson, Jr. Comprehensive survey of artificial intelligence—the study of how machines (computers) can be made to act intelligently. Includes introductory and advanced material. Extensive notes updating the main text. 132 black-and-white illustrations. 512pp. 5⅜ × 8½. 24864-X Pa. $8.95

HISTORY OF INDIAN AND INDONESIAN ART, Ananda K. Coomaraswamy. Over 400 illustrations illuminate classic study of Indian art from earliest Harappa finds to early 20th century. Provides philosophical, religious and social insights. 304pp. 6⅜ × 9⅜. 25005-9 Pa. $8.95

THE GOLEM, Gustav Meyrink. Most famous supernatural novel in modern European literature, set in Ghetto of Old Prague around 1890. Compelling story of mystical experiences, strange transformations, profound terror. 13 black-and-white illustrations. 224pp. 5⅜ × 8½. (Available in U.S. only) 25025-3 Pa. $5.95

ARMADALE, Wilkie Collins. Third great mystery novel by the author of *The Woman in White* and *The Moonstone*. Original magazine version with 40 illustrations. 597pp. 5⅜ × 8½. 23429-0 Pa. $9.95

PICTORIAL ENCYCLOPEDIA OF HISTORIC ARCHITECTURAL PLANS, DETAILS AND ELEMENTS: With 1,880 Line Drawings of Arches, Domes, Doorways, Facades, Gables, Windows, etc., John Theodore Haneman. Sourcebook of inspiration for architects, designers, others. Bibliography. Captions. 141pp. 9 × 12. 24605-1 Pa. $6.95

BENCHLEY LOST AND FOUND, Robert Benchley. Finest humor from early 30's, about pet peeves, child psychologists, post office and others. Mostly unavailable elsewhere. 73 illustrations by Peter Arno and others. 183pp. 5⅜ × 8½. 22410-4 Pa. $3.95

ERTÉ GRAPHICS, Erté. Collection of striking color graphics: *Seasons, Alphabet, Numerals, Aces* and *Precious Stones*. 50 plates, including 4 on covers. 48pp. 9⅜ × 12¼. 23580-7 Pa. $6.95

THE JOURNAL OF HENRY D. THOREAU, edited by Bradford Torrey, F. H. Allen. Complete reprinting of 14 volumes, 1837–61, over two million words; the sourcebooks for *Walden*, etc. Definitive. All original sketches, plus 75 photographs. 1,804pp. 8½ × 12¼. 20312-3, 20313-1 Cloth., Two-vol. set $80.00

CASTLES: THEIR CONSTRUCTION AND HISTORY, Sidney Toy. Traces castle development from ancient roots. Nearly 200 photographs and drawings illustrate moats, keeps, baileys, many other features. Caernarvon, Dover Castles, Hadrian's Wall, Tower of London, dozens more. 256pp. 5⅜ × 8¼. 24898-4 Pa. $5.95

AMERICAN CLIPPER SHIPS: 1833–1858, Octavius T. Howe & Frederick C. Matthews. Fully-illustrated, encyclopedic review of 352 clipper ships from the period of America's greatest maritime supremacy. Introduction. 109 halftones. 5 black-and-white line illustrations. Index. Total of 928pp. 5⅜ × 8½.
25115-2, 25116-0 Pa., Two-vol. set $17.90

TOWARDS A NEW ARCHITECTURE, Le Corbusier. Pioneering manifesto by great architect, near legendary founder of "International School." Technical and aesthetic theories, views on industry, economics, relation of form to function, "mass-production spirit," much more. Profusely illustrated. Unabridged translation of 13th French edition. Introduction by Frederick Etchells. 320pp. 6⅛ × 9¼. (Available in U.S. only)
25023-7 Pa. $8.95

THE BOOK OF KELLS, edited by Blanche Cirker. Inexpensive collection of 32 full-color, full-page plates from the greatest illuminated manuscript of the Middle Ages, painstakingly reproduced from rare facsimile edition. Publisher's Note. Captions. 32pp. 9⅜ × 12¼.
24345-1 Pa. $4.95

BEST SCIENCE FICTION STORIES OF H. G. WELLS, H. G. Wells. Full novel The Invisible Man, plus 17 short stories: "The Crystal Egg," "Aepyornis Island," "The Strange Orchid," etc. 303pp. 5⅜ × 8½. (Available in U.S. only)
21531-8 Pa. $4.95

AMERICAN SAILING SHIPS: Their Plans and History, Charles G. Davis. Photos, construction details of schooners, frigates, clippers, other sailcraft of 18th to early 20th centuries—plus entertaining discourse on design, rigging, nautical lore, much more. 137 black-and-white illustrations. 240pp. 6⅛ × 9¼.
24658-2 Pa. $5.95

ENTERTAINING MATHEMATICAL PUZZLES, Martin Gardner. Selection of author's favorite conundrums involving arithmetic, money, speed, etc., with lively commentary. Complete solutions. 112pp. 5⅜ × 8½. 25211-6 Pa. $2.95

THE WILL TO BELIEVE, HUMAN IMMORTALITY, William James. Two books bound together. Effect of irrational on logical, and arguments for human immortality. 402pp. 5⅜ × 8½. 20291-7 Pa. $7.50

THE HAUNTED MONASTERY and THE CHINESE MAZE MURDERS, Robert Van Gulik. 2 full novels by Van Gulik continue adventures of Judge Dee and his companions. An evil Taoist monastery, seemingly supernatural events; overgrown topiary maze that hides strange crimes. Set in 7th-century China. 27 illustrations. 328pp. 5⅜ × 8½. 23502-5 Pa. $5.95

CELEBRATED CASES OF JUDGE DEE (DEE GOONG AN), translated by Robert Van Gulik. Authentic 18th-century Chinese detective novel; Dee and associates solve three interlocked cases. Led to Van Gulik's own stories with same characters. Extensive introduction. 9 illustrations. 237pp. 5⅜ × 8½.
23337-5 Pa. $4.95

Prices subject to change without notice.
Available at your book dealer or write for free catalog to Dept. GI, Dover Publications, Inc., 31 East 2nd St., Mineola, N.Y. 11501. Dover publishes more than 175 books each year on science, elementary and advanced mathematics, biology, music, art, literary history, social sciences and other areas.